DESIGN & MANAGEMENT
For RESOURCE RECOVERY
Volume 2
HIGH TECHNOLOGY—
A FAILURE ANALYSIS

Ragman—resource recovery at the turn of the century.
Courtesy of the National Center for Resource Recovery

DESIGN & MANAGEMENT
For RESOURCE RECOVERY

Volume 2
HIGH TECHNOLOGY—
A FAILURE ANALYSIS

by
Brenda Harrison
P. Aarne Vesilind

P. Aarne Vesilind
Series Editor

ANN ARBOR SCIENCE
PUBLISHERS INC / THE BUTTERWORTH GROUP

TD
794
.5
·H36

SERIES PREFACE

Given today's economic, social and political indicators, pessimism seems to be the path of least resistance. The United States and the rest of the western world are already importing most of the requisite raw materials to feed the industrial state. Nations that own these resources, recognizing the economics of scarcity, will charge ever-increasing prices for them, thus fueling worldwide inflation. In a material- and energy-scarce society, Keynesian economics no longer apply, and inflation, unemployment and a steadily decreasing standard of living seems to be part of our future.

Much of this pessimism, however, is based on the assumption that the resource use cycle is one-way and open-ended—materials are extracted, used and discarded, to be lost forever. Fortunately, this is not necessarily a correct assumption, for it is possible to reclaim useful materials from waste and inject these back into the resource cycle. Likewise, much of the organic fraction in wastes, which derives its energy from the inexhaustable energy source—the sun—can be burned to produce useful heat or processed to yield other fuels.

The science and technology associated with such a reclamation of materials and energy from waste has become known as resource recovery. It is a fledging technology, borrowing unabashedly from other fields such as mining engineering, biochemistry, chemical process engineering and thermal sciences. Although it has experienced growing pains (often as the result of overenthusiasm and inadequate technical and managerial skills), it nevertheless seems to be an absolute requisite for an orderly transition through the 1980s and into the 21st century.

The series *Design and Management in Resource Recovery* is the first continuing series of books devoted to this topic. The authors all have wide-ranging knowledge in their individual areas of expertise and have produced volumes which are both contemporary and useful as reference works for years to come.

Reading and editing the series is not only a rewarding experience, but has substantially abated my feelings of pessimism about the future. Learning to design and manage resource recovery systems will not guarantee us a happy and prosperous future, but it is at least one meaningful bridge that will help bring it about.

P. Aarne Vesilind

PREFACE

In the spring of 1979, the unfortunate state of affairs in the still-young resource recovery industry was depressing. Millions of dollars had been committed, with great expectations and good intentions, and there was little cause for rejoicing. Few high-technology resource recovery facilities had been even marginally successful.

It was obvious that one way of benefiting from such apparent fiascos was by learning from the mistakes. This was difficult, however, because some of the facts behind the failures were obscured, either intentionally or otherwise, and were not readily available to the public. We decided that it would be a challenge to investigate some of the failed projects where the valuable lessons to be learned were not readily evident. The five systems discussed in Chapters 2 to 6 of this book were selected to provide a representative cross section of high-technology resource recovery attempts, all of which were at one time considered failures, although some of which ultimately proved to be modest successes. These five projects represented different technologies, and the factors contributing to the projects' failures were also different. Research was undertaken through literature reviews, personal and telephone interviews with individuals associated with each system, and plant visits when possible. The information-gathering phase was conducted by the senior author while she was enrolled as a graduate student in the environmental engineering program at Duke University.

It is our hope that our original objective—to provide historical perspective and factual information and thus increase the knowledge in the field of resource recovery—has been accomplished. We feel the book can be most beneficial as a tool to be used by municipal officials and planners when they approach the question of resource recovery/solid waste disposal in their own communities. Armed with a realistic outlook and the experiences of previous approaches to high-technology resource recovery, they can build on the experiences of their colleagues a decade ago and thus avoid the potential pitfalls that can beset a community starting a recovery program.

We are indebted to the Richard C. Leach Endowment Fund which, through the Duke School of Engineering, enabled us to conduct the research necessary for this book. We are also indebted to municipal and public officials; state and federal environmental protection agency staff members; plant operators and former plant project managers; legal and engineering consultants; professors; and many other contacts too numerous to name who freely gave time and information during the research for this book. Finally, the senior author wishes to express special thanks to her parents for their unfailing support and encouragement. It is to Wilbur and Jean Harrison that this book is dedicated.

Brenda L. Harrison
P. Aarne Vesilind

Brenda L. Harrison is a consultant in solid and hazardous waste management with the environmental consulting firm of Gershman, Brickner & Bratton, Inc. Ms. Harrison received an MS in Environmental Engineering and a BS in Civil Engineering from Duke University.

Her work has included technical and economic resource recovery feasibility analyses for various municipalities and visits to and research of various resource recovery systems. She has also co-authored an article on modular incineration for *Waste Age* magazine.

P. Aarne Vesilind is Associate Professor of Civil Engineering at Duke University, Durham, North Carolina, and Director of the Duke Environmental Center. Born in Estonia, he holds BS and MS degrees in civil engineering from Lehigh University, and a Master's degree and a PhD in sanitary engineering from the University of North Carolina. Dr. Vesilind has also been a Fellow at the Norwegian Institute for Water Research in Oslo, and recently was the recipient of a Fulbright-Hayes Senior Lectureship to study in New Zealand.

Among his many honors and awards are the Duke University Outstanding Professor Award for 1971-1972 and the 1971 Collingwood Prize from the American Society of Civil Engineers.

Dr. Vesilind is the author of two Ann Arbor Science Publishers, Inc. publications, *Environmental Pollution and Control* and *Treatment and Disposal of Wastewater Sludges*.

CONTENTS

CHAPTER 1

THE EMERGENCE OF HIGH-TECHNOLOGY RESOURCE RECOVERY

The reclamation of materials and energy from solid waste is an old and time-honored profession. In all organized societies the waste products have contained items which, although valueless to the original owner, became a source of income for the scavenger. The ragman of the 19th century was such an integral part of cities that he was immortalized in song and fable.

Today, the secondary materials industry is a multimillion dollar business. Scrap steel, aluminum, glass, etc., are all recovered and recycled in vast quantities. With few exceptions, however, this industry is concerned only with industrial scrap. Mixed municipal refuse has been considered to be too "dirty" and unpredictable to allow for the recovery of its specific fractions.

In the early 1970s, however, conditions changed sufficiently in a number of ways to force a closer evaluation of reclaiming materials from municipal solid waste. First, more and more communities were experiencing difficulties in developing solid waste disposal schemes. Due to the ease of individual transportation, land close to cities was being developed for residential use, thus becoming very expensive. Further, there was increased public resistance to the siting of landfills. Solid waste was, as one frustrated mayor described it, "what everyone wants picked up but no one wants put down" [1]. Likewise, incinerators were becoming increasingly expensive due to added costs of air pollution control. Many cities (New York and Los Angeles, for example) even began to close down their existing incinerators due to the anticipated costs of stricter air pollution requirements. As a result, city planners and engineers began to study alternative means of disposing of their solid waste.

The second factor influencing the growth of modern scavengering was the realization of our dependence on materials and energy. We as a nation could no longer be fully independent if we had to depend on the good will of others to supply the raw materials for our industry and energy production.

Several reports [2-4] published by the National Academy of Sciences and others have confirmed this fear, and the oil embargo brought it home to the public. The reclamation of energy and materials from waste thus became a useful political standard and a popular public concern.

Thirdly, the idea of recovering the valuable components from solid wastes was enhanced indirectly by the so called "Oregon Bottle Laws." These laws, patterned after the pioneering legislation in Oregon, are intended to reduce beverage container litter by placing a mandatory deposit on all beverage containers. This gives returnable containers an artificial economic advantage and forces the bottlers to change to the returnable containers. The great fear of the various industries associated with beverage container manufacturing, as well as national breweries and soft drink firms that have developed regional production and marketing strategies based on the one-way containers, was that the Oregon law would become national. Although substantial energy savings would be realized as the result of such legislation [5] and a substantial reduction in litter achieved [6], the short-term dislocation of the brewing/bottling/container manufacturing industries would be severe.

Hoping to avoid this, the various interested parties funded a public relations and involvement organization (Keep America Beautiful Inc.) and began to promote the idea of recovering materials and energy from solid waste as a constructive alternative to restrictive legislation. The latter effort resulted in the formation of the National Center for Resource Recovery in Washington, a nonlobbying, technology and policy transfer operation. This organization, by working with such established organizations as the American Society for Testing and Materials (ASTM), the U.S. Environmental Protection Agency (EPA), the U.S. Department of Energy (DOE) and others, has been providing the impetus, skill and leadership in promoting resource recovery. In fact, the name "resource recovery" was developed by the group that eventually became the Center.

Three forces—fewer and more expensive alternatives for refuse disposal; material and energy shortages; and the promotion of resource recovery by industry as a constructive alternative to restrictive legislation—all converged in the early 1970s to produce an era of euphoria and expectation of great things to come.

It soon became evident that one of the most economical means for reclaiming the energy and materials from such a heterogeneous mixture as municipal solid waste (MSW) was to develop sophisticated technology that could be applied to large facilities. Many of these first resource recovery facilities in the United States were thus "high-technology facilities," meaning that machinery and/or mechanization was employed in processing the waste stream. This contrasts with low-technology options such as hand picking through refuse, used years ago in some cities in the United States, and which is still an integral part of solid waste processing in many developing countries.

For example, from 1898-1901 the New York City Department of Sanitation reclaimed secondary materials by hand sorting, yielding an income of $1/ton [7]. The main problem confronting the application of high technology to refuse processing was that most of these unit operations were either new or had never before been shown to work with municipal refuse. The Resource Recovery Act of 1970 was designed in part to spur the application of high technology to resource recovery by providing funds for full-scale demonstration projects. Concurrently, some cities and a few private firms decided to plunge into the resource recovery business without the benefit of federal assistance. Thus, during the early 1970s many high-technology projects were initiated, all in an atmosphere of "gold rush" optimism.

Some of these projects never developed any impetus, while others became operational but at a much higher cost than originally expected. None could be considered an unqualified success story.

The reasons for these "failures" have been numerous and varied, including, but not limited to, inexperience with an untried or unproven technology, insufficient quantity of waste input, lack of markets for recovered products, unforeseen inflationary trends, and incomplete or inadequate cost projections for the projects. Both public and private sectors alike have paid dearly for the increased knowledge in the field of resource recovery.

Consequently, views toward recovering energy and/or materials from the solid waste stream have changed. Citizens, community officials, private corporations and private investors have become wary of the optimistic claims of resource recovery systems that were often heard in the late 1960s and early 1970s. More and more, communities want to be assured that the chosen recovery system is the least expensive and most beneficial (an energy source for the area, for example) disposal method for their particular area. They want the risk factor associated with the project's success to be as small as possible. These views are somewhat mirrored in the Resource Conservation and Recovery Act of 1976 (RCRA), which reduced the emphasis on testing new recovery technologies and focused more on the planning behind solid waste utilization and disposal decisions.

There is every reason to believe that high-technology resource recovery facilities will continue to develop as fast as in the past decade, however. Landfills, typically the least expensive form of waste disposal, are becoming harder to locate and more expensive to operate. This fact, coupled with the rising cost and scarcity of energy, often makes a solid waste recovery plant competitive with other disposal methods for MSW if energy recovery is included in the facility plans.

If the chances of success in planned or future high-technology resource recovery facilities are to be improved, the past experiences of similar plants should be critically reviewed. Perhaps the failures and successes in the planning, design, construction and operational stages of these facilities (as judged by hindsight) can be avoided or included, respectively, in future endeavors. Of course, and as mentioned throughout the resource recovery literature, factors affecting the ultimate financial outcome of each facility are quite typically site-specific, excluding the technical aspects of the facility design. Keeping this in mind, much can be learned from the past decade's growth in resource recovery.

It is generally recognized by those in the resource recovery field that not all of the data or facts behind some of the recovery plants have been publicly disclosed. Even with the recent experiences with high-technology resource recovery facilities, application of past experience with respect to future projects could be unfeasible without access to particular information. Under these circumstances, knowledge valuable to planners of future recovery systems may be overlooked.

For instance, private firms with proprietary resource recovery systems and processes frequently gloss over or fail to volunteer all of the information regarding their operation and its relative success or failure. Concern over public opinion may influence those in charge of municipally run facilities not to disclose all of the facts about their operations or financial situations. Failure to mention potentially embarrassing facts or events involving a party associated with a recovery facility may also occur. On the other hand, information behind a facility may be available, but it requires time and effort to collect it from a wide range of sources. Leads necessary to find these sources of information are not always apparent.

To add to the wealth of information regarding high-technology resource recovery, the planning, design and operation of five representative projects are presented in this book. These facilities are: (1) the Nashville Thermal Transfer Corporation plant; (2) the St. Louis Union Electric solid waste utilization system proposal; (3) the Baltimore pyrolysis system; (4) the Lowell incinerator residue project; and (5) the San Diego flash pyrolysis system. These five systems were chosen because the full story of each seems incomplete and because each has been considered a "failure" at some point in time, i.e., each failed to meet original technical or financial expectations. Of the chosen facilities, only the Nashville system was not partially funded through the EPA demonstration grant program.

The objectives of this book are threefold:

1. to thoroughly investigate the planning, design and operation of each of the five facilities;
2. to document the information gained from the investigation so past events in resource recovery can be viewed more realistically; and
3. to develop common denominators of success or failure that may be evident from investigation of the five facilities.

Each of the recovery facilities is discussed in one of the following chapters (2-6). Although the complete picture of each system is given, areas of common knowledge for a particular system—such as information covered in EPA reports—are mentioned only briefly. Emphasis is placed on defining the factors that led to the failure of each facility to meet design expectations.

REFERENCES

1. Welch, L., Mayor of Houston Texas. In: *The Problem of Solid Waste Disposal,* E. A. Glysson, J. R. Packard and C.H. Barnes, Eds. Ann Arbor, MI: University of Michigan Press.
2. "Requirements for Fulfilling a National Materials Policy," *Proc. Engineering Foundation Conference,* Henniker, NH, Office of Technical Assignment, Washington, DC (1974).
3. National Commission on Materials Policy, "Materials Needs and the Environment—Today and Tomorrow," Washington, DC (1973).
4. Committee on Mineral Science and Technology, "Mineral Science and Technology—Needs, Challenges and Opportunities," final report, NTS, Washington, DC (1969).
5. Bingham, T. H. "The Economic and Energy Impact of a National Container Deposit Law," Research Triangle Institute, Research Triangle Park, NC (1976).
6. Fenner, T. W., and R. J. Gorin. "Local Beverage Container Laws," Stanford Environmental Law Society, Stanford University Law School (1976).
7. Hering, R., and S. A. Greeley. *Collection and Disposal of Municipal Refuse.* (New York: McGraw-Hill Book Co., 1921).

CHAPTER 2

THE NASHVILLE THERMAL
TRANSFER CORPORATION

There is no magic black box where you can stick in garbage and out comes
steam energy.

—Ben McDermott*

Whether it be steam energy or liquid fuel oil, Nashville was one of the first
communities in the United States to learn that the production of energy
from solid waste is not a simple matter. Nor is it an inexpensive one.

The Nashville Thermal Transfer Corporation (NTTC or Thermal) has been
incinerating MSW to produce steam for the heating and cooling of downtown
Nashville buildings since 1974. The first few years, however, were stormy,
and the project was considered a failure by many. As one Nashville public
works official stated, "It [NTTC] has weathered political, legal, technical,
and financial barriers that many observers felt would be fatal" [1]. Claims of
near bankruptcy, periods of burning waste only at night to hide the resultant
air pollution problems, and lack of support by Nashville citizens all contributed to this popular opinion. Today, however, these concerns seem to
have faded and Thermal appears to be a viable resource recovery project.
Nevertheless, by our definition, initially it was a "failure."

---

*As quoted in *The Nashville Banner*, April 20, 1977. Mr. McDermott was the general
manager of Nashville Thermal Transfer Corporation at that time.

BACKGROUND INFORMATION

The idea for Thermal came as an answer to a combination of needs in Nashville and Davidson County during the late 1960s, when heating and cooling requirements were coupled with severe solid waste disposal problems. The Metropolitan Government of Nashville and Davidson County (METRO) developed plans for a centrally located, conventionally fueled heating and cooling plant to provide services to the city's municipal buildings. The engineering firm conducting the study, I. C. Thomasson and Associates, Inc., suggested a larger system encompassing more of the downtown area, taking advantage of an upcoming urban renewal project that included placement of utility, communication, water and sewer lines beneath the streets [2]. The size and scope of the proposal were found to be economically feasible, especially if the network of heating and cooling pipes was to be installed while the streets were under excavation during the renewal stage [3].

It was first expected that local gas and electric utilities might be responsive to managing such an undertaking; however, the city's charter does not allow such a proposal. An amendment to change the charter would have required a referendum—an inherently lengthy procedure. With the economics of the facility related to the timing of the urban renewal construction and to the ability to quickly provide services to large downtown buildings then under construction, a timely go-ahead for such a centralized heating and cooling facility was considered a priority [4].

The city's law director and bond counsel determined the best alternative to a utility-owned facility would be the establishment of a nonprofit public corporation having the ability to issue tax-exempt revenue bonds. Provisions for such an institution are found under the Internal Revenue Service 63-20 ruling, which allows bonds issued from a corporation formed under a state's nonprofit corporation law to be tax-exempt under the provisions of section 103(a)(1) of the Internal Revenue Code of 1954, provided that the five requirements stated in the ruling (63-20) are met. Thermal met these requirements. By May 14, 1970, the Nashville Thermal Transfer Corporation was in existence. Stipulations for the corporation included that it carry out a public service; that its directors be public officials; that the easements for the piping networks be given to it free of charge; and that its assets be given to the city on debt retirement. Initial estimates for the facility were given as $4.5 million. As financing for debt repayment and operational expenses was to be obtained through revenues to the facility, the city quickly instituted a search for potential customers.

At the same time, METRO was having increasing problems with its waste disposal. Its five landfills were open dumps, described to have been in "pretty grim" condition [5]. Legislation passed by the state required all landfills to use cover materials by 1972 [6], but cover material was expensive and almost impossible to find because the geology of the county is largely near-surface limestone bedrock [7]. Additionally, new sites were scarce and difficult to acquire.* To make matters worse, METRO citizens were showing "an unprecedented talent for generating solid waste," approaching 5.5 lb/capita/day [7].

The idea to incinerate solid waste instead of using fossil fuel to produce the steam for the district system, combining two pressing needs in the area, consequently was met with approval.** Although such a project was novel for this country, similar usages of steam produced from incinerated wastes were known to be operating in Europe. Mayor Beverly Briley strongly supported the plan, and I. C. Thomasson & Associates, Inc. was commissioned to prepare a new feasibility study for such a venture. The study projected additional plant costs of $8.4 million, but felt the ultimate savings in fossil fuel costs and waste disposal costs warranted the resource recovery facility. Additionally, the plant was determined to be technically feasible [4]. From this point on, the "cash for trash"+ project was on its way.

FACILITY EXPECTATIONS AND INITIAL DESIGN

The incineration of solid waste (mostly MSW) in the Thermal facility was to be mutually beneficial to the city of Nashville, to Thermal and to the 29 downtown buildings that were customers of the facility. The city would be decreasing its solid waste collection and disposal costs by transporting waste to a centralized facility that was expected to have a zero-charge dumping fee. A newspaper article published in *The Tennessean* on November 7, 1973, went so far as to claim that METRO would save approximately $1.25 million per year in decreased transportation costs alone. Thermal would be getting zero-cost fuel for steam production. Buildings serviced by the centralized plant were projected to have a 25% decrease in heating and cooling costs when compared to separate, inhouse systems [3].

The facility was designed to use mass-burning, water-wall incineration,

---

*A 1979 estimate gives only 0.5% of the metropolitan land area to be available and suitable for landfill, with these areas being along the river and in the floodplain. Existing landfills were projected at having a 0.5- to 3-year life [5].

**It is not definite as to who initially thought of this idea. Some sources claim it was I. C. Thomasson, some say both I. C. Thomasson and Mayor Briley, others say it was Adrian Gammill, a waste consultant for the Thomasson engineering firm.

+This phrase, commonly used in reference to the Thermal facility, was coined by Maurice Wilson, design consultant with the Thomasson engineering firm (*The Nashville Banner*, September 16, 1971).

similar to systems frequently found in Europe. Technology and equipment were American, however. Additionally, incineration of MSW for cooling purposes had never been attempted in European applications. Although the first phase of the project involved a 650 metric ton/day (720 ton/day) capacity, provisions for future expansion possibilities (such as onsite layout area and in sizing of original steam and cool water piping) were considered.

Figure 2-1 shows a simplified flow diagram of the system. Unprocessed waste was to be gravity fed to two 325 metric ton/day (360 ton/day) Babcock & Wilcox Company water-well boilers housing Detroit Stoker Company, four-level, reverse reciprocating grates. Continuous steam output of 24 hr/day, 365 day/yr, was to be 480,000 N/hr (109,000 lb/hr) at 323°C (613°F) and 2760 kPa (400 psig) [8]. Both boilers were to be capable of using either oil or gas to maintain combustion stability of efficiency. A standby, Combustion Engineering Company fossil fuel-fired boiler was also provided for backup purposes. Figure 2-2 shows the plant as completed.

Steam was designed to be distributed at 1033 kPa (150 psig) in the heating network, with resultant condensate returned to the facility. For cooling purposes, steam-driven turbines operating two Carrier Corporation centrifugal

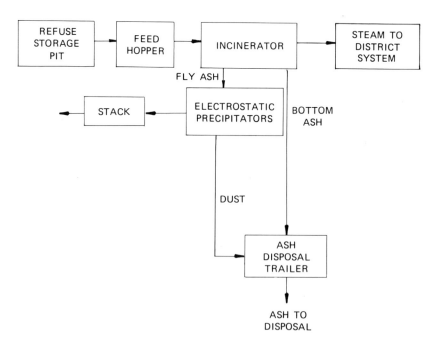

Figure 2-1. Nashville thermal operational schematic, as originally designed.

Figure 2-2. The Nashville Thermal Transfer Facility as it looks today.

chillers were to produce water of 4°C (40°F). Water of approximately 13°C (56°F) would be returned to Thermal [9].

The inert residue from the incinerators was projected to be sterile enough to not require the added expense of landfill cover. Additionally, it was thought that the overall air quality of Nashville would be improved through the use of a centralized facility, as opposed to separate systems for each customer building [3]. Possibilities of recovering ferrous materials from the incinerator residue, thereby further reducing the amount of waste to be land-filled, were to be considered in the future.

Lastly, the Nashville Thermal plant would be located downtown, near its customers, as is necessary for steam district heating and cooling. The facility was to be aesthetically pleasing, blending in with the downtown area.

INITIAL FUNDING

As it turned out, the initial funding available for the Thermal facility was the most important factor contributing to the early failures of the recovery plant to meet initial goals. To offer revenue bonds to raise project design, construction, contingency and operational funds, Thermal had to earn support from the financial community. Several requirements to assure project viability were set by local investment bankers and the NTTC nine-member Board of Directors. Most importantly, Thermal needed commitments (contracts) from potential customers "to insure above break-even operations" [4]. These contracts had to include escalation clauses that would allow rate increases, if necessary, so Thermal could meet its debt repayment schedules. Additionally, an unbiased economic evaluation of the plant was found to be necessary.

The first customer contract was obtained in June 1970 with the state of Tennessee for 30 years of service to several of its large buildings in the downtown area, such as the state capitol. Next, METRO committed itself to a 30-year agreement for its area buildings. Included in this contract were provisions that METRO retain ownership of the waste, deliver it to the district facility and then transport and dispose of plant residue for the 30-year period. Thermal would not charge dump fees and the city would not charge for waste delivery [6].

Private customers remained hesitant in committing themselves to a contract, however, until a $2 million line of credit granted to Thermal by a local consortium of banks further convinced them of the project's viability. The loan enabled NTTC to obtain a downtown site for the plant, thereby allowing the project to remain on schedule for a projected 1972 construction start [4].

It should be obvious that while private owners with buildings under construction might not have to construct inhouse heating and cooling systems, such a measure would put a building owner in a vulnerable position should Thermal have difficulties in providing heating and cooling services on schedule or in a consistent fashion. The Nashville Hyatt Regency Hotel, for instance, is presently fully dependent on Thermal for such services, having no backup systems.

Based on estimates and pre-bids on major equipment items, it was projected in 1972 that a total of $16.5 million in bond issues would be required. Included in this estimate were plant costs of $8.4 million and distribution costs of $4.0 million [10]. Revenues to back only about a $14 million bond offering had been established in customer contracts, however [11]. Thermal went back to the city to negotiate a small dumping fee that resulted in an additional $150,000 annual payment to the corporation. Additionally, attorneys for NTTC were able to arrange for annual pseudorevenue to the facility through legislation to the state that exempted Thermal from paying property taxes. This decrease in expenses was argued as income to the corporation, in effect, and was used to reach the necessary annual revenue level to back a $16.5 million bond issue [11].

After all the preparation, however, bids on construction costs for the facility as originally designed were significantly higher than the anticipated $12.4 million. McEwen and Levy [10] claim that the bids were $3.5-$5 million greater than expected.

Thermal was then faced with having to decrease construction expenditures and/or increase project funding. The decision was made to cut costs and modify the design of the plant until total project costs corresponded to the $16.5 projection. It is not known whether initial attempts to increase funding were also made during this period, however. At some point before construction started, a Nashville contingency approached the EPA for funding assistance. Evidently, they were "turned down flat" and told the plant was "proven technology" [12]. NTTC was ultimately able to decrease costs to the $16.5 million level. As discussed in a subsequent section, such cost-cutting measures were responsible for most of the problems in the Nashville incinerator facility.

Both the state and the metropolitan governments also wanted NTTC to obtain a tentative permit for the facility from EPA before bond issuance [4]. One of the cost-cutting measures had involved the replacement of the original design for electrostatic precipitators by a combination of mechanical collectors and low-energy wet scrubbers; however, EPA expressed to both the state and city its feeling that this form of pollution control would not meet state and federal air pollution standards [13]. With no operating evidence of this specific technology on mass incinerators, however, no real proof could be given that the less expensive control devices would not work [11]. The EPA permit system is such that the agency has the final say regarding a system's compliance with standards only *after* a system is operating; therefore, EPA's hands were tied.

A permit from the local health department was all that was actually necessary to begin construction. The system designers were finally able to convince the METRO Board of Health to issue a tentative permit for the facility when a performance guarantee from the manufacturers of the low-energy scrubber (Air Conditioning Corporation) was obtained [4].

Thermal at last was ready to offer $16.5 million worth of revenue bonds to the market, with an economic feasibility study of the Heating and Cooling Revenue Bonds by the Duff and Phelps, Inc. engineering firm contained in the offering statement. A negotiated sale using a syndicate of area investment bankers was the chosen method of financing, with a resulting average interest rate of 5.7%. The bonds were then sold in a matter of days in the market to twelve large institutions [4].

Two interesting factors should be noted. First, Resource Planning Associates [4] claim that "one of the principals responsible for the underwriting was not even aware that EPA had expressed any reservations about the system's performance." It has not been determined if the other financiers were also lacking such knowledge. Second, the project was marketed as a proven system when selling the bonds [11]. This, however, was not an accurate assessment of the facility, especially with respect to the air pollution devices used for the system.

PROJECT OPERATION

The operational history of Nashville Thermal can be divided into two rather distinct periods. The first period, termed the "failure period" for this discussion, was directly linked to the effects of having to work with insufficient funds, as mentioned previously. Although some of the technical and financial problems that occurred were to be expected in normal shakedown of any large undertaking such as an incinerator plant, the major technical problems were the result of design or equipment decisions made so that total project costs could meet the budgeted $16.5 million raised in revenue bonds. The second period, or "recovery period," began approximately when additional funding was secured. This financing enabled Thermal to begin correcting most of the technical problems, paying for outstanding debts and covering operational expenses. Thermal now appears to be at the end of this second period.

The Failure Period

Construction of the cash-for-trash plant began in July 1972—almost immediately after the initial funding was secured—and lasted until January 1975. In an effort to obtain additional revenues, however, operations began roughly a year before the end of construction to provide heating and cooling services to several customers [14]. An additional incentive of an early start was the need to provide service to customers who had no in-house systems [10]. The first three to four months of these services were accomplished using the fossil fuel—fired package boiler, as the waste-burning boilers were still under construction. Beginning operation expenses were consequently high because allowances had not been made for the purchase of such large amounts of fossil fuels. Since the incineration of solid waste in June 1974, however, Thermal has been burning mainly the zero-cost fuel. Over the time span of June 1974 to July 1976, 68% of the energy was provided by municipal solid waste [9].

Testing in July 1974 of the low-energy wet scrubbers used for air pollution control officially revealed major air pollution problems. The particulate readings of 0.41 grains per dry standard cubic foot of air (gr/dscf) corrected to 12% carbon dioxide (CO_2) were roughly five times higher than the allowed federal particulate standards for incinerators of 0.08 gr/dscf @ 12% CO_2 [15].

Meanwhile, the citizens of Nashville did not need to know actual particulate readings to realize Thermal was polluting the atmosphere—large amounts of black smoke coming from Thermal's stacks beginning June 1974 were evidence enough. Thermal consequently began burning wastes only at night, using gas as fuel during the day. The press, which had optimistically publicized Thermal's benefits until this point, commented in *The Tennessean* on July 3, 1974 [16]:

> The present method—emitting clouds of smoke under cover of darkness—may have merit in terms of public relations, but is highly questionable in terms of true public interest.

The METRO Board of Health extended Thermal's tentative permit while the facility tried to improve the scrubbers' performances. It is reported that the board had ordered Thermal closed in December 1974, although it reversed its decision by the end of the month [16].

The attempts to improve the collection abilities of the low-energy scrubbers, such as increasing the fan speed, adding venturi rods and changing some operational procedures, were only partially successful. The most efficient operation gave emissions of 0.1685 gr/dscf @ 12% CO_2 in February 1975 [15]. The EPA's previous doubts regarding the capabilities of the low-energy scrubber for the Nashville project were obviously correct.

In addition to being unable to meet the municipal waste incinerator standards, the scrubbers were experiencing severe corrosion and erosion problems [10]. Thermal finally realized it would have to totally replace the scrubbers with a more efficient emission control unit so that standards could be met and both the press and the citizens of Nashville could be placated. A local engineering firm, which had been working on increasing scrubber performance, recommended an electrostatic precipitator (ESP) for one boiler and a testing program of two baghouse pilot units on the other boiler. If either baghouse unit were successful, Thermal would install such a system on its second boiler since the total costs were expected to be lower than with an ESP. If neither were successful, an ESP would be used for the second boiler [14].

An EPA compliance order was issued in the spring of 1975, requiring Thermal to purchase and install an electrostatic precipitator, test the baghouse systems and install either an ESP or baghouse as determined by the tests. Dates were set for fulfillment of these requirements. Additionally, Thermal was required to retrofit the current air pollution system during the interim period before installing new devices [15]. Obviously, expenses of Thermal continued to mount, as the cost of retrofitting a problematic piece of equipment and of buying and installing new pollution devices were soon to be added to the expense of purchasing large amounts of fossil fuels.

Another major technical problem experienced at Nashville Thermal was the deterioration of the lower water-wall boiler tubes beginning in the fall of 1974, ultimately resulting in both boilers being disabled in the spring of 1975. Replacement of all these tubes was found to be necessary. These failures have been attributed to the absence of refractory lining on the tubes, insufficient combustion air throughout the incinerator and an accumulation of material on the tubes themselves, increasing the rate of corrosion [14].

Significantly, most of the factors contributing to the tube wastage were the result of the last-minute austerity program before construction. The boiler manufacturers convinced the engineering firm that a silicon carbide refractory for the lower tubes—a design feature used in many water-wall designs—was not really necessary, thereby decreasing the boiler costs [12]. (The original plans had included specifications for the refractory.) The tubes were consequently much more subject to the corrosion and erosion influences of burning heterogeneous MSW. Further, accumulation of the particulates could have been much more easily dislodged from such a refractory-coated tube than from one without such a covering.

The insufficient combustion air was partially a result of inadequate over-fire air in the incinerator [12], although it is not known whether this deletion was also made to reduce costs. Insufficient overfire air would have created a predominant underfire air supply, which, in turn, would have increased the particulate matter entrained in the gas stream. With more particulates in the gas stream, the possibilities of erosion and corrosion would increase. Moreover, insufficient overfire air led to insufficient particle burn-out [12].

Other cost-cutting choices affected project success, although not as significantly as the previously discussed decisions. Use of steam-driven turbines to provide energy for much of the auxiliary plant equipment was used to decrease operational expenses. However, lack of operating flexibility and some unexpected shutdowns of the boilers resulted from this dependence on the steam production [9]. Control and feedback mechanisms were also originally at a minimum to cut plant costs [17], thereby decreasing possibilities of adequate operation of the incinerators.

Not all of the problems occurred as a result of attempts to reduce total facility costs, however. For instance, water return from the distribution system was plagued with high hardness, necessitating expensive water treatment. Stoppage of the leakages into the piping network that had been creating the water quality problems required much time and effort because some of the inputs originated from the customer's side [18]. Problems with unwanted vibration in one of the chillers also required correction [10].

As these technical problems surfaced, operational and maintenance (O&M) expenses immediately began to climb. Total costs were not offset by revenues, especially because customer demands did not develop as projected [19].

By February 1975 NTTC had announced a significant rate hike for its services. Still another rate increase was instituted three months later in May 1975. Thermal evidently invoked the first increase without knowing the extra amount of revenues needed. The second increase was set after the amount of needed revenues was known [11]. The total of both increases gave approximately a 100% increase in revenues [14].

With two successive rate increases and ongoing pollution problems, Thermal was not conveying a good image to the public. Besides the poor public relations, however, Thermal was in financial trouble. As of May 31, 1975, Thermal had a construction overrun of $2.2 million, an operating deficit of $3.1 million and proposed construction requirements through 1980 to total $4.1 million [10].

Rumors of bankruptcy began with coverage by local newspapers. One article spoke of an Arthur Anderson & Company financial audit that felt NTTC needed new financing by 1976 to pay its debts. Evidently, Thermal had not met certain bond requirements, claiming bankers could legally foreclose on the corporation at any time [20]. METRO attorneys established that METRO would have no legal responsibility if such an event occurred, although it was realized that investors had viewed the project as "an arm of the Metropolitan government" [21].

The Recovery Period

Thermal was able to escape most of its financial and, consequently, its technical problems when funding of approximately $9.5 million above the original $16.5 million bond issue was made available. Briefly, $2.3 million was secured through a bond issue purchased by three local banks; $5.7 million was obtained through a state loan; and up to $1.5 million per year was promised by the Metropolitan Council [22]. Acquisition of this funding was not simple, however, as some of the early attempts—such as introducing another bond issue of $8 million to the market—were met with opposition [23].

The installation of two American Air Filter Company electrostatic precipitators was ultimately completed during this stage of Thermal's life. Resultant particulate emissions since have been easily within the limits of 0.08 gr/dscf @ 12% CO_2. Stack emissions are now invisible to the untrained eye.

Both of the waste-fired incinerators were also retrofitted during this period. Silicon carbide coatings were put on all the lower water-wall tubes, thereby updating them to state-of-the-art technology. It is not known how much of this repair, if any, was paid for by Thermal. Chambliss [12] claimed the boiler manufacturers came back and put the refractory lining on at their own expense, although this "required a lot of arm twisting" by the I. C. Thomasson firm. (As mentioned before, the manufacturers had evidently convinced the engineering firm that the refractory lining was not necessary.) The superheater units were replaced with larger, more durable versions [9]. Additionally, the original gravity waste feed system to the boilers did not provide sufficient waste feed control. Recent additions of hydraulic ram feeders have allowed much more control over the waste feedrate—an important operational variable in incineration performance [17]. As a result of these and other modifications, each boiler has a maximum rated capacity of 480 metric ton/day (530 ton/day), 150 metric tons (170 tons) above the original capacity [6].

With air pollution problems alleviated and boiler operation improved, the proportions of waste feed used as fuel steadily increased. For example, during the winter of 1976-77, when fossil fuels were not abundant, Thermal easily continued operating using waste as fuel. By 1977 and 1978 Thermal had reached the point of using wastes for 94% of its steam production [6].

Financially, Thermal finally was in good standing. By the end of fiscal year 1978, for instance, Thermal only needed a $1.3 million payment from METRO and the ratings on Thermal's bonds improved [6].

CURRENT STATUS

Thermal is presently incinerating an average of 360–380 metric ton/day (400–420 ton/day), or roughly one-fourth of the waste generated in the METRO area. Use of each boiler is alternated to accomplish periodic maintenance on each unit, such as grate repair and replacement [17]. Only during times of maximum demand will both incinerators be used concurrently. It is not expected that current throughputs will be increased [5] because economics are based on the use of solid waste as fuel and the provision of continuous and reliable services to Thermal's customers. Continuous use of both boilers to increase throughput is not practical because unexpected or lengthy maintenance on one or both of the boilers would leave only the high-cost, fossil fuel–fired boiler for steam production.

With all the modifications, however, problems could still arise. Milton Kirkpatrick, manager of Thermal, has stated that the underground piping network "is our big worry" if a break in the pipes should occur. However, he feels that Thermal "could handle just about any problem that could come up in the plant" [24]. The corporation is also stocking spare parts for maintenance purposes, something it could not do in the past because not enough capital was available [17].

It has not been established whether METRO is actually saving money because of the Thermal operation. Heidenrich [5] expressed that the the Department of Public Works at least had found it beneficial in decreasing accidents and wear on equipment. An annual METRO payment of $1.3 million would convert to roughly $9.25/metric ton ($8.40/ton) for disposal before including residue disposal costs (assuming the plant can average 408 metric ton/day (425 ton/day). The department is responsible for disposing of roughly 113 metric ton/day (125 ton/day) of residue (incoming waste is reduced roughly 75% by weight and 90% by volume). Residue is usually sterile, although times of very high-moisture MSW may result in residue that is "warmed over" [5]. Tennessee is therefore requiring at least final landfill cover when disposing of the residue.

Customers of Thermal are presently enjoying consistent service. Heating and cooling costs are based on two factors: (1) a set or minimum fee for being serviced by the corporation; and (2) a usage fee for the services by unit price. As usage increases, the unit price of the steam and cool water decreases [6]. Although present prices have been said to be the same or higher than prices for services through conventional means, it is expected that increases in fossil fuels, coupled with savings from not installing in-house systems, will make Thermal prices at least comparable, if not better, than conventional service prices.

EVALUATION

It appears that Thermal has recovered from its stormy beginnings. Although some of the original expectations—such as the provision of low-cost heating and cooling services or the provisions of zero-cost waste disposal—have not been realized, NTTC currently *is* operating. MSW is being environmentally disposed of; steam and cold water are consistently being produced; and the expensive and scarce landfill areas are being used at a slower pace.

For the most part, lack of sufficient financing led to many of Thermal's early difficulties. Had enough been available, the two major technical problems of the deterioration of the water-wall tubes and the inability to remove enough particulates to meet air pollution standards would not have occurred. Additionally, some of the equipment, design and construction "short-cuts" could have been eliminated.

As with any new undertaking, it was difficult to correctly estimate costs for Thermal's operation and maintenance. Avers, a past Thermal manager, spoke of these uncertainties [25]:

> This [Thermal] was a new project, the first of its kind. There weren't ten other people we could go talk to about costs. Our estimates were low. We now know rather precisely what our expenses will be.

Had more money been available initially, however, the cost overruns due to unexpected operational costs and problems would not have been so great. The importance of adequately planning and providing for high beginning operational costs cannot be overstated.

That the initial inability to generate adequate funding was allowed to have a detrimental effect on the cash-for-trash project points to two factors. First, the time–pressure element created a situation in which decisions had to be made quickly to implement the district heating and cooling process. As mentioned before, it was thought a priority that the waste incinerators be constructed during the downtown renewal project and in time to attract potential customers. This time–pressure factor was responsible for creating the not-for-profit NTTC. It also was influential in agreeing to use low-energy scrubbers as soon as a performance guarantee was obtained.

Second, the absence of a controlling body which had ultimate responsibility for NTTC seemed apparent during the decision-making processes for Thermal before its construction [10]. Many groups had inputs into the as-constructed design, but no one group seemed accountable for these inputs and final decisions. Such a group should have required an unbiased technical evaluation of the project. It appears this was never done. An outside engineering firm only focused on the financial aspects of the project, using previously estimated costs and expenses [4]. Such a responsible body might also have paid more attention to the strong EPA warning about the feasibility of the scrubber system for air pollution control.

The facts that at least one investment banker did not know of the EPA warning and that the bonds were sold as backing a proven technology show, at the least, some lack of communication between the groups having input into the project. Had the doubts about the scrubbers been more widely known, the Thermal project may not have been able to begin without provisions for electrostatic precipitators for pollution control.

A few points should be mentioned regarding air pollution standards and technology at the time of the design of the Thermal plant. Chambliss [12] believes that the uncertainty regarding ultimate air emission requirements, in addition to the desire to decrease costs, was a factor in the decision to use the scrubbers. Evidently, it was speculated that EPA might require gaseous pollutant control as well as particulate control for newly constructed waste incinerators, and the I. C. Thomasson firm thought the scrubbers would be able to remove both forms of pollutant. Electrostatic precipitators, on the other hand, can only remove particulates.

However, with the adoption of the New Source Performance Standards (NSPS), any incinerator burning at least 50% MSW and built, altered or under construction by August 17, 1971, was subject only to particulate standards (0.08 gr/dscf @ 12% CO_2). Additionally, data were available in

1971 to show that ESP's, baghouses and high-energy scrubbers could meet the particulate standards [26]. Low-energy scrubbers were not mentioned. Significantly, construction did not begin until the middle of 1972.

The performance guarantee obtained for the scrubber claimed a 95% removal of particulates having a mean diameter of 5μ [9]. Such a guarantee was accepted without determining actual particle size distribution of the particulates from the Thermal boiler. However, the three previously mentioned devices capable of reaching the 0.08 standards have high collection efficiencies for particles much smaller than 5μ, suggesting that such capabilities are necessary to meet standards. As it turned out, emission tests of the boilers in 1975 showed up to 40% of the particulates to be below the $5\text{-}\mu$ range [8,9]. By design, then, the low-energy scrubbers are not technically capable of removing the necessary amounts and sizes of particulates for a waste-fired incinerator. It is unfortunate that Thermal ended up learning this fact through first-hand experience.

It is to Thermal's credit that it was able to come out from under the "black cloud" that seemed to cover it during its early years of operation. This was due in no small part to the abilities of the 40-man staff to adapt and modify operating procedures to obtain optimum results and to assist with required modifications [14]. Further, the Metropolitan government's support was integral in helping put the Thermal facility on the road to recovery.

To summarize, the incineration of solid waste to produce steam for district heating and cooling is a technically feasible undertaking, as long as correct, state-of-the-art equipment is used. In Nashville's case, the combination of scarce and costly landfill, the need for a district system, and an urban renewal project seemed to point to an economically attractive project. Whether such a project would be economically feasible for a specific locality would depend on site-specific factors such as alternative disposal costs, local energy costs and available markets for the heating and cooling services. Nashville's experience has undoubtedly helped change the "cash for trash" or "black box" attitude to energy recovery from waste to a more realistic and conservative view.

REFERENCES

1. Heidenreich, P. "Nashville Thermal Transfer Corporation, *APWA Reporter* 7-8 (1975).
2. Centec Corporation. "Executive Briefing Report: Waste-to-Energy Systems," draft copy of report for Environmental Resource Information Center, U.S. EPA, Contract No. 68-03-2672 (1979).
3. Briley, B. "We Will Air Condition and Heat with Garbage," *Am. City* 61-62 (1972).

4. Resource Planning Associates. Financial Methods for Solid Waste Facilities, Environmental Protection Agency, PB 234-612 (Springfield, VA: National Technical Information Service, 1974), pp. 264-269.
5. Heidenreich, P., Assistant Director, Department of Public Works, Nashville. Personal communication (1979).
6. Bernheisel, J. F. "Nashville: A Successful Refuse-to-Energy Program," *NCRR Bull.* 14-17 (1979).
7. Heidenreich, P. "Nashville Develops a Solid Waste System," *Solid Waste Syst.*, 3-7 (1974).
8. Bozeka, C. G. "Nashville Incinerator Performance Tests," *Proc. 1976 NWPC*, Boston, MA (1976), pp. 215-27.
9. Ralph M. Parsons Company. *Engineering and Economic Analysis of Waste to Energy Systems,* Environmental Protection Publication 600/7-78-086 (Springfield, VA: National Technical Information Service, 1978), pp. 45-61.
10. McEwen, L. B., and S. J. Levy. "Can Nashville's Story be Placed in Perspective?" *Solid Wastes Management* (1976).
11. Stokes, O., former attorney for Metro Law Department, Nashville. Personal communication (1979).
12. Chambliss, C. W., associate with I. C. Thomasson Associates, Inc., Nashville. Personal communication (1979).
13. Hellwig, V., air enforcement branch for EPA Region IV office, Personal communication (1979).
14. Avers, C. E. "Technical-Economic Problems in Energy Recovery Incineration," *Proc. 1976 Nat. Waste Processing Conf.,* Boston, MA (1976), pp. 59-66.
15. Sauer, R. E. "Electrostatic Precipitator Cuts Back Refuse Incinerator's Air Pollution," *Solid Wastes Management* 56:51-52 (1979).
16. Bauman, K. "Nashville or Trashville: Which Will it Be?" Unpublished results, Duke University (1975).
17. Dugan, M., NTTC control room operator, Personal communication (1979).
18. Patton, S. R., P. M. Houston and W. R. Gattman, "They're Making a Successful Trash-Energy Swap in Nashville," *Resource Recovery Energy Rev.* 22-23 (Fall 1977).
19. Avers, C. E. "Nashville Confronts Technical, Economic Problems in Refuse-to-Energy," *Prof. Eng.* 45 (11):32 (1975).
20. *Nashville Banner* (September 17, 1975).
21. *The Tennessean* (November 31, 1975).
22. "Nashville Thermal Back on Track, Proves Value During Big Freeze," *Solid Waste Syst.*, 8-9 (1977).
23. *The Tennessean* (March 10, 1976).
24. York, M. "Life is Cooling Off at the Thermal Plant," *The Tennessean* 66-68 (1977).
25. *Nashville Banner* (April 16, 1975).
26. U.S. EPA. "Background Information for Proposed New Source Performance Standards," Research Triangle Park, NC (1971).

CHAPTER 3

THE UNION ELECTRIC SOLID WASTE UTILIZATION SYSTEM

... I think we can make it [The Union Electric proposal] pay—if we can be competitive in price and convenience with the landfill, and if we can get enough trash.

—Gerald Rimmel*

In February 1979 the city of St. Louis reluctantly realized it could not make any version of the Union Electric (UE), 6530 metric-ton/day (7200 ton/day) cocombustion system "pay." It was felt, in fact, that the project would not be competitive with area landfill costs when utilizing the realistically obtainable waste quantities of 3600 metric-ton/day (4000 ton/day) in the area [1].

Plans for the UE Solid Waste Utilization System (SWUS) had been originally proposed in 1974 as an outgrowth of a 270 metric-ton/day (300 ton/day) EPA demonstration facility cofinanced by both the city and UE. Although the demonstration project was generally considered a success, major delays and problems with the SWUS proposal caused UE to conclude in 1977 that it could no longer finance and manage the project. By this time the project engineering design was roughly 85% complete [1] and the city had become dependent on the proposal as a way out of its growing waste disposal needs [2].

The plans for the SWUS were then passed to a state development agency

---

*As quoted in the *St. Louis Globe—Democrat*, June 20, 1978. Mr. Rimmel is on the Bi-State Development Agency staff.

in hopes that the agency could instigate and manage such a system. UE indicated it could still use the fuel produced from the municipally run plant. However, two years of studies ultimately showed the plan to be unfeasible and the city subsequently abandoned the entire project.

The St. Louis story is really two stories; a success followed by a failure. The successful demonstration of a system was not sufficient to ensure the implementation of a full-scale program.

ST. LOUIS DEMONSTRATION PLANT

The St. Louis project was the first system ever to demonstrate combustion of processed MSW with coal in a utility boiler. Financed by the EPA, UE and the city of St. Louis, the 270 metric-ton/day (300 ton/day) facility operated intermittently from July 1972 to November 1975.

The processing of MSW into a refuse-derived fuel (RDF) took place at the refuse processing facility located at the site of the city's southern incinerator. As shown in Figure 3-1, waste was fed by conveyor to a 1250-hp shredder, where it was reduced to roughly 2.5-cm (1 in.) particle sizes. The shredded waste was then conveyed to a rectangular air classifier for separation into a light, RDF fraction and a heavy, metal-laden mostly inorganic fraction. Ferrous materials were removed from the heavies via a magnetic belt, with the rejects disposed of in a landfill. The RDF fraction was conveyed to a 1000-m³ (35,000 ft³) storage bin to await the upcoming 30-km (18 mi) trip to the UE boilers via city trucks (Figure 3-2).

At the Meramec power plant, RDF was deposited in a receiving bin and ultimately transported to a storage facility (Figure 3-3). From this point, pneumatic pipelines transferred the RDF to a 125-MW Combustion Engineering, tangentially fired, suspension-type boiler, which had been modified to accept the RDF by replacing gas burners from each corner of the boiler with RDF-firing ports. Refuse was successfully fired with pulverized coal in proportions of 5–20% by heating value [3].

Some operational problems occurred in the demonstration facility but for the most part these were solved. High particulate emissions from the air classifier cyclone indicated a need for some type of pollutant control (such as a baghouse) for a future facility. High attrition rates in the carbon steel elbows of the pneumatic pipelines transporting the RDF were noted. A successful study was subsequently undertaken to locate more wear-resistant piping materials [4].

Not surprisingly, bottom ash quantities in the boiler increased when firing RDF. Other significant operational changes included an increased particulate loading and gas flowrate to the existing electrostatic precipitator, decreasing its efficiency [3].

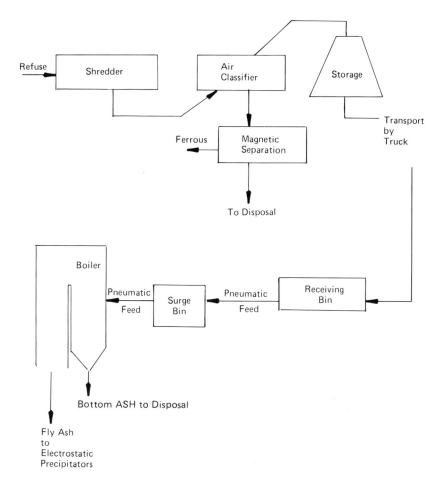

Figure 3-1. Union Electric Solid Waste Utilization System pilot plant operational schematic.

THE UNION ELECTRIC PROPOSAL:

Initial Plans

The Union Electric Company viewed the demonstration facility to be a successful operation. Even before facility termination in 1975, UE announced its plans to build, own and operate a $70 million, 7200 metric-ton/day (8000 ton/day), 6 day/wk Solid Waste Utilization System through a wholly owned subsidiary called the Union Colliery Corporation (UCC). The system

Figure 3-2. The St. Louis solid waste processing facility. The shredder is on the right; air classifier tower is in the middle; and the storage shed is on the left of the photograph.

Figure 3-3. The Union Electric power station where RDF was test-burned. The RDF storage silo is in center left.

was to be patterned after the demonstration project and set up in conjunction with an onsite metals processor for ferrous recovery [5].

To collect the necessary wastes, an 11,600-km² (4500 mi²) area spanning the city of St. Louis and surrounding areas was to be covered. UE initially proposed a collection/transport system consisting of the construction and operation of five strategically located transfer stations of 1350–1800 metric-ton/day (1500–2000 ton/day) capacities. Once at the transfer station, up to 900 metric-ton/day (1000 ton/day) of waste would be railed to the UE Meramec plant and up to 5400 metric-ton/day (6000 ton/day) railed to the UE Labadie plant, each system using existing, but separate, railroads. Processing plants of nominal 1800 and 5400 metric-ton/day (2000 and 6000 ton/day) capacity would be constructed at the Meramec and Labadie plants, respectively. A June 1977 date was set for operation.

Revenue to support the construction and operation of such a large-scale private undertaking was to come from tipping fees from area wastes and sales of recovered metals and RDF. No measures were to be taken to instigate wasteflow control. Instead, UE planned to attract the waste tonnages by offering a competitive tipping fee. Additionally, UE claimed that "no money to build or operate the system will come from Union Electric's electric customers" [5].

Refinements to the initial plans later decreased total tonnages to be processed to 6480 metric-ton/day (7200 ton/day). Such a reduced tonnage, still the largest ever to be proposed for a waste processing facility, was to be transported, processed and cofired entirely at the Labadie site, roughly 60 km (40 mi) west of St. Louis. Additionally, the number of transfer stations planned was reduced to four facilities. Two different rail lines were still to be used, increasing assurance of a steady fuel supply. Figure 3-4 shows a flow diagram of the proposed collection/transportation network. UE was also able to set up another onsite firm to process the heavy fraction from the air classifier for nonferrous metals, such as aluminum. Rejects from this facility ultimately would be returned to the processing plant.

Project Delays and Problems

From the initial SWUS proposal in February 1974 to UE's decision to terminate project implementation in February 1977, the project was beset with problems.

Perhaps the most infamous setback encountered by UE was the difficulty in obtaining a site for its southwest transfer station. Area residents were determined that such a facility would not be built in the vicinity, and several court battles ensued. A local town ordinance prohibiting "dumping or reduction of solid waste, garbage or offal" was used for the defense, although UE

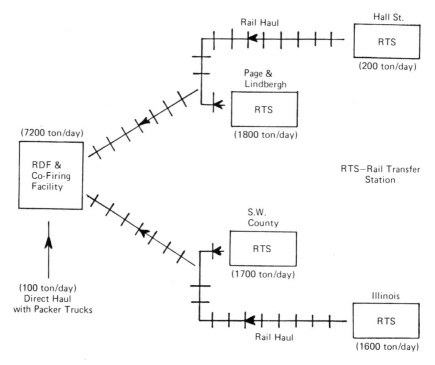

Figure 3-4. Flow diagram of collection and transportation for the proposed full-scale St. Louis Union Electric Solid Waste Utilization System. [1]

maintained that transfer of solid waste from one container to another did not violate the ordinance [6].

Since the fourth transfer station was integral to the overall SWUS, UE placed the project on hold in September 1975 to wait for the legal decision. Waste equipment manufacturers were asked to postpone shop fabrication of units designated for the facility, and already processed machinery was kept in storage. Meanwhile, inflationary factors inevitably began to take hold because of such a delay.

Although the court did decide the case in the utility's favor, a subsequent request by the town to repeal this decision was accepted in February 1977. In addition, the Appeals Court was unable to immediately set a date for the hearing [6]. A further delay in project implementation seemed imminent, along with a rise in system costs.

Two events developed that had a synergistic-like effect when combined with the transfer station siting delays. According to one source, the state of Missouri enacted legislation prohibiting public utilities from borrowing money for interest on construction projects [7]. Even more important, Proposition 1 was approved in November 1976 by Missouri voters. This law prevents public utilities from adding costs of new facilities to customer rates before said facility is operational [8]. With a nuclear plant under construction since 1976 [7], UE was obviously facing cash flow problems.

In February 1977, after the court decision to reconsider the transfer station disagreement, UE officials terminated the project, even though $14 million had already been spent [6]. C. J. Doughtery, President of UE, was quoted as saying [8]:

> We regret the necessity for this announcement. However, the fact is that there is a limit to the ability of a business to accomplish economic and desirable ends when controlling decisions are made on non-business grounds.

As an interesting footnote, the Missouri Court of Appeals ultimately decided in favor of the town on January 16, 1979 [9]. It should also be pointed out that UE had claimed originally that no costs from the SWUS would be incurred by its customers. The reaction of UE to the passage of Proposition 1 indicates a discrepancy with earlier claims.

IN BI-STATE'S HANDS

On termination of the SWUS by UE, the city of St. Louis was suddenly faced with no alternative for solving its solid waste problems. The two city-owned incinerators were inadequate, needing expansion and upgrading.

In 1966 the city had obtained its contribution to the demonstration project from part of a $3.37 million general obligation bond originally earmarked for needed expansion and increased pollutant control of its two 450 metric-ton/day (500 ton/day) waste incinerators. These expansion and improvement plans had been postponed, however, while participating in the cocombustion project [2].

Determined to convert refuse to RDF for cocombustion in Union Electric's power plant, officials such as the city mayor and the county supervisor immediately approached the Bi-State Development Agency (BSDA) to see whether the agency would consider implementing the project [10]. BSDA, a regional organization of the city of St. Louis and six surrounding counties in Illinois and Missouri, has the power to plan, implement and operate projects in these areas, such as the conveyor system in the Gateway Arch.

With BSDA's abilities to issue revenue bonds, consideration was given to arranging, financing and managing much of the system, with ultimate sale of the RDF to UE. Engineering design of SWUS was already 85% complete, and land for three of four transfer stations had been secured.

With the project in Bi-State's hands, the agency initially turned to EPA for assistance in evaluating such a program. This was done because BSDA felt it did not have enough experience in solid waste to consider such a decision on its own. However, Bi-State did realize two important facts from the beginning–the area "had a lot of landfills and 80–90% of our [the area's] trash was carried by private haulers" [10]. It was felt that acquisition of 6500 metric-tons/day (7200 ton/day) would be difficult, possibly requiring waste flow control legislation. Therefore, the Agency asked EPA to look at 3600 and 1800 metric-ton/day (4000 and 2000 ton/day) systems, in addition to the UE-proposed 7200 ton/day plan. By December 1977, a report entitled "Evaluation of the Union Electric Proposed Resource Recovery Project" was presented to BSDA by the EPA's Office of Solid Waste [11].

Opinions regarding the usefulness of the report have been generally negative. J. Eigner, who was hired by BSDA as project manager for the UE proposal, expressed that BSDA and others were "not thrilled" by the report. He felt the evaluation was "not a very in-depth study," as they took "UE's numbers and updated them a little bit, but they didn't really dig in." Evidently, "they just accepted very old numbers" [10].

The conclusions in the report were not definite, especially with respect to financial predictions. However, some main points from the evaluation seemed to be useful. First, EPA emphasized the need for assurance of waste delivery for the project's success, indicating doubts that 6500 metric-ton/day (7200 ton/day) could be secured realistically. Second, it was determined that the onsite ferrous metals processor set up in the initial plans for SWUS did not feel it was economically feasible to build the planned detinning plant at capacities below 3600 metric-ton/day (4000 ton/day), indicating some economic dependence on the large-scale system. Third, a sensitivity analysis conducted for the project indicated waste throughput, O&M costs and RDF value to be the most critical factors impacting cost estimates developed by UE. In reaching these sensitivity conclusions, however, EPA basically used operation and maintenance, capital and revenue prices as provided by UE. As discussed later, actual costs were found to deviate substantially from original UE projections. A final estimate was given for necessary funding behind a 3600 metric-ton/day (4000 ton/day) system (this being an $81.6 million bond issue for a system of $70 million in capital costs).

EPA also provided information regarding UE interaction with a system implemented through BSDA. Significantly, a letter to the EPA team from D. L. Klumb on July 18, 1978 included the fact that the BSDA system would have to produce RDF "in conformance to UE specifications" [12]. Further, UE felt it would probably reject a system management arrangement. Obviously, the management of a large facility by BSDA would involve much greater risk to UE than a situation in which UE makes and uses the product, as originally planned.

All in all, BSDA was unsatisfied with the EPA report. The Agency consequently went to the city and county requesting expert advice on the feasibility of implementing a 3600 metric-ton/day (4000 ton/day) version of the SWUS. Both government bodies contributed a substantial sum for the hiring of a financial advisor and ultimately a waste management consultant. This study commenced in September 1978.

Meanwhile, citizens and community officials were not happy with the delays in Bi-State's decision regarding the Union Electric proposal. Perhaps most vocal was J. Shea, the city commissioner of refuse, who was faced with the operation of two inadequate incinerators. In June 1978 he criticized the delay by saying, "for the last year or so, the plan [SWUS] has been sitting in the lap of the Bi-State Development Agency." With still no decision by January 1979 he was quoted as saying, "We don't need more studies. We've studied this until we're blue in the face" [13].

Shea did not have long to wait, however, as BSDA received the waste management consultant's report in February 1979 [1]. Briefly, the report declared any version of the SWUS to be economically unfeasible, citing high disposal costs for 1982 of $14-26/ton to be the deciding factor. Such high disposal costs would not be competitive with estimated 1982 landfill costs of $6.60/ton. Further, project financing through revenue bonds would be almost impossible without waste flow control legislation. The area haulers group had given the SWUS plan a conditional endorsement in 1978, but such an endorsement stipulated low dumping fees and no waste flow control [14]. The total bond issue for a 3600 metric-ton/day (4000 ton/day) facility system was estimated as ranging from $154-$161 million, almost two times higher than the previous EPA estimate. Such a great difference in estimates was attributed to large increases in both equipment and O&M costs; more up-to-date cost information on operating large RDF facilities; the reluctance of material processors to construct new facilities with only 3600 metric-ton/day (4000 ton/day) capacity plant; and the need for a more detailed and costly collection network [1]. This study was more in-depth and specific regarding economics—almost all cost inputs to the system were separately costed-out through manufacturers, systems designers, etc.

Two points of the study need elaboration. First, UE had factored in revenues from charging onsite metal processors land leasing and railroad switchyard fees, besides selling them the unprocessed metals. With the absence of these processors at a 3600 metric-ton/day (4000 ton/day) facility, these revenues would not exist.

Second, the management consultants' plan called for a more complicated transportation/collection network than originally planned. This was justified by the reasoning that convenience was important to the waste hauler. If transfer stations could be located near major landfills or hauler garage facilities, he or she would be more likely to deliver wastes to the system. On looking at the locations of the landfills, the garages and the solid waste generation areas, the consultants concluded it would have to provide several transfer stations to capture 3600 metric-ton/day (4000 ton/day) of waste. Interestingly, the plan included the conversion of both city incinerators to transfer stations. The added scope of the transportation/collection system consequently increased overall costs. Significantly, operation and maintenance costs of the transportation system alone were projected to be greater than 50% of the entire system O&M costs.

The ultimate recommendation was that BSDA abandon the SWUS plan and consider future installations that would be smaller and more centrally located. These waste-to-energy systems could thus better satisfy the waste disposal needs of the city and surrounding areas. In February 1979 BSDA terminated all attempts to implement a regional solid waste system that would produce RDF for cofiring in Union Electric pulverized coal boilers.

EVALUATION

One of the most significant realizations from the experiences in St. Louis is that the Solid Waste Utilization System proposal was thought to be *technically* viable. All postponements and doubts in the development of the large-scale facility were *financially*, not technically, motivated.

It would be misleading to claim that the technology of cocombustion has no unsettled technical problems, however. Other systems have been subsequently constructed using the technology originally demonstrated at the 270 metric-ton/day (300 ton/day) plant, although none can yet be considered an overwhelming success. Questions have not been totally answered regarding such factors as possible boiler operation problems with long-term cocombustion or air emission control problems.

However, it appears that these unanswered questions were not thought to be important enough to stop Union Electric from originally proposing such a large-scale system. In fact, adequate redundancy had originally been scheduled into the plant by providing an entire standby processing line. UE considered the technical problems solved, and financial reasons forced it to back down from its proposal. The utility has continued to express interest in buying the RDF.

Even more significantly, an important fact clearly shown in the St. Louis experience is that site-specific factors are the foundation of a resource recovery/solid waste processing facility. If one of these foundations is faulty, it needs to be repaired if a facility is to remain on a stable footing. Such repairs could include additional money for adequate transportation/collection systems or legislation for waste flow control. After-the-fact attempts at adjusting the site-specific factors are frequently impossible or too costly, as in St. Louis. A less expensive and more successful approach would be to match the facility with the site-specific factors. This will be attempted in the near future if Bi-State initiates some smaller, more centrally located energy recovery facilities. Waste flow control would not be a problem with a small facility as city-collected MSW in the city boundaries could be routed to provide waste tonnages. Additionally, the transportation costs could be kept at a minimum with more centrally located plants. The fact that BSDA would have had to construct and operate such a large and complex collection/ transportation system increased projects costs greatly.

Local availability and cost of landfill are the most important site-specific criteria influencing a waste facility's economic success. Much of the area surrounding St. Louis had adequate and relatively inexpensive (as compared to the ultimate SWUS tipping fees) landfills. Additionally, the large number of private haulers created a competitive collection business [1] in which customers could frequently choose the least expensive hauler. High tipping fees would not attract the private haulers, then, unless *no* landfills or less expensive means of disposal were available. Significantly, the tipping fees predicted in the most recent study for the 3600 metric-ton/day (4000 ton/ day) system would have been competitive with disposal costs in areas where landfills are expensive and scarce.

The failure of the large-scale Union Electric solid waste facility to be implemented in St. Louis means only that such a system is not the correct choice for that area. Had UE been able to avoid the initial financial and siting problems it encountered originally, it would still have had to deal with the perhaps insurmountable difficulties in obtaining such large daily waste tonnages.

The idea of cocombustion with Union Electric has finally been purged from the thoughts of city officials and citizens. Although "the only ultimate solution" to the waste disposal problem "is to eat the paper bag with the hamburger" [15], the city of St. Louis can now take a more objective approach to finding a solution to its solid waste disposal problems.

REFERENCES

1. DSI Resource Systems Group, Inc. "A Feasibility Analysis for the Implementation of a Regional Resource Recovery System in the St. Louis Area, Vol. 1, Executive Summary," prepared for Bi-State Development Agency; and "A Feasibility Analysis for the Implementation of a Regional Resource Recovery System in the St. Louis Area, Vol. 2, Final Report," prepared for Bi-State Development Agency (1979).

2. Resource Planning Associates. *Financial Methods for Solid Waste Facilities*, Environmental Protection Publication PB 234-612 (Springfield, VA: National Technical Information Service, 1974).
3. Fiscus, D. E., P. G. Gorman, M. P. Schrag and L. J. Shannon. "December 1977, St. Louis Demonstration Final Report: Power Plant Equipment, Facilities and Environmental Evaluations," Environmental Protection Publication EPA-600/2-77-155 a and b (Springfield, VA: National Technical Information Service, 1977).
4. Murphy, J. D. "Materials to Resist the Abrasion of Pneumatically Transported Processed Refuse" *Proc. 8th ASME NSWPC*, Chicago, IL (1978).
5. Klumb, D. L. *Union Electric Company Solid Waste System*, 2nd ed., St. Louis, MO (1955).
6. "Union Electric Abandons $70-million Wastes Utilization Project in St. Louis," *Solid Wastes Management* 3:54,76 (1977).
7. Klumb, D. L. In: "St. Louis," by J. Downing, Duke University, Unpublished results (1978).
8. "Union Electric Gives up Waste Fuel Plan," *Am. City Country* 4:13,16 (1977).
9. *St. Louis Post-Dispatch* (January 18, 1979).
10. Eigner, J., Director of Solid Waste Project Management, Bi-State Development Agency, St. Louis, Missouri. Personal communication (1979).
11. Office of Solid Waste, USEPA, "Evaluation of the Union Electric Proposed Resource Recovery Project," prepared for Bi-State Development Agency (1977).
12. Klumb, D. L., UE manager of SWUS project. Personal communication (1979).
13. *St. Louis Globe–Democrat* (June 20, 1978; January 12, 1979).
14. *St. Louis Globe–Democrat* (June 20, 1978).
15. Shea, J. In: *St. Louis Globe–Democrat* (June 18, 1978).

CHAPTER 4

THE BALTIMORE CITY PYROLYSIS DEMONSTRATION

It's easy for them [Monsanto] to walk away. The city is still here.

—Mayor William Donald Schaefer*

Since his inauguration in December 1971, the mayor of Baltimore has not been known to mince words. The above quote was one of his more conservative comments when Monsanto Enviro-Chem Systems, Inc., designer and partial funder of the Baltimore pyrolysis demonstration facility, terminated its involvement with the project in February 1977.

As originally designed, the plant was expected to process and pyrolize 900 metric-ton/day (1000 ton/day) of MSW, resulting in salable steam, ferrous materials and glassy aggregate. Waste by-products either were to be recycled or landfilled. Since the beginning of operation in January 1975, however, many technical problems emerged. Shutdowns were frequent and money for modifications was liberally spent. Eventually, Monsanto felt that further modifications would still not be able to reach initial project performance guarantees as stipulated in 1973, and since it had almost reached the limits of its budget on the experimental facility, it withdrew from the project.

The fact that the city "was still there" is the underlying reason behind the present operational status of the facility today. Baltimore had participated in the demonstration facility under the assumption that the project would solve successfully its waste disposal problems. Faced with

---

*As quoted in the *Baltimore Evening Sun*, February 3, 1977.

a potential problem when Monsanto dropped project responsibility, the city of Baltimore was forced to attempt to make the project (and its investment) a success. Although the plant probably will never reach initial throughput or economic expectations, as of November 1979 it was processing 540 metric-ton/day (600 ton/day) of MSW and producing (and selling) roughly 1.3-1.5 million lb/day of steam [1].

Baltimore has been perhaps the most infamous of the initial resource recovery projects that failed to meet initial expectations.

BACKGROUND INFORMATION

Monsanto began research in 1967 on solid waste problems, concentrating on disposal because of experience in materials processing. Pyrolysis was felt to be one of the most attractive disposal methods, and a direct-fire pyrolysis using a rotary kiln was the chosen method. After working with a laboratory-size model, a 32 metric ton/day (35 ton/day) pilot plant was built in St. Louis. By 1970, continuous operation of the system was obtained; and by 1971, processes to recover ferrous material, char and glassy aggregate had been added. A scaleup to the 900 metric-ton/day (1000 ton/day) system was thought to be reasonable, as similar scaleup ratios in both petrochemical and materials processing industries were typical [2].

Concurrently, EPA was looking for demonstration technologies, and the Monsanto Landgard® system was thought to have an excellent chance of being successful. In fact, the Deputy Assistant Administrator of Solid Waste Management Programs in EPA at that time commented that of all the demonstration projects funded by EPA, the Monsanto project was thought to have the highest chance of success [3].

Meanwhile, Baltimore was faced with growing amounts of waste, plus incinerators requiring repair and upgrading to meet federal standards. Mayor Schaefer indicated that unless the public works officials made progress in alleviating the solid waste problem by his inauguration, "... they won't be around here anymore" [4]. The head of the Public Works Department, Linaweaver, saw the St. Louis demonstration and convinced the newly elected mayor to back the project.

However, several officials did express doubts regarding the pyrolysis system before it was accepted. In a 1972 letter to Linaweaver, the head of the Bureau of Engineers expressed that untried and unproven methods (such as the pyrolysis proposal) should not be used because the need for additional solid waste reduction facilities was urgent. He recommended that an incinerator be built instead [5].

With EPA and the Director of Public Works pushing for the pyrolysis facility, Baltimore decided to go with the demonstration. Initial costs were estimated as $14.6 million: $6 million contributed by the EPA; $4 million from the Maryland Environmental Service (MES); and $4.6 million from the city. A turnkey contract was settled, with a provision for $4 million in performance penalties to be paid by Monsanto if all air emission standards were not met; if plant capacity could not reach an 85% average capacity for an identified 60-day period; or if the putrescible content of the residue were greater than 0.2% [2].

FACILITY DESIGN AND EXPECTATIONS

The Baltimore pyrolysis plant was expected to process 900 metric-ton/day (1000 ton/day) of MSW, recovering roughly 4.8×10^6 lb of steam, 63 metric-tons (70 tons) of ferrous metal, 152 metric-tons (169 tons) of glassy aggregate, and 72 metric-tons (80 tons) of carbon char (or solid residue) daily. The glass product was to be used in "glassphalt" as an aggregate for road building, and a market for the ferrous fraction had been established. The char was to require landfilling unless a market could be found for it [6].

The low-grade steam ultimately produced from the combustion of the gaseous pyrolysis product was to be transported via pipeline to the Baltimore Gas and Electric Company (BG&E), less than a mile from the site in southern Baltimore. As with Nashville Thermal, the close proximity of the steam market to the facility is necessary for economic transportation of the product. A five-year contract was established, with prices based on a percentage of costs of No. 6 fuel oil. Steam pressure was to be 690-1800 kPa (100-260 psig), with temperatures less than 213°C (415°F) [2].

Original plant economics were estimated to give net costs of $0.02/ton of waste by February 1974. Such low costs were related to the estimates of steam revenues associated with the fuel oil price increase in 1974 [2].

The process flow diagram in Figure 4-1 shows the original design of the facility. Briefly, waste would be weighed and deposited in a pit, conveyed to either shredder, then transferred to either a storage bin or the rotary kiln. The shredding was to be 10 hr/day. The storage bin allowed a 24 hr/day, 7 day/wk operation of the remaining units.

The pyrolysis process was to begin with continuous waste feed via hydraulic rams to the refractory-lined 30-meter (100 ft)-long, rotary kiln. Three zones would be established for the waste—a drying zone, a pyrolysis zone and a combustion zone. Volatile matter would be vaporized, with the resultant gases flowing countercurrently to the incoming wastes and out into the gas purifier. The inert residue would continue out the other end, being quenched in a tank and directed to a materials recovery system.

It should be mentioned that the term "pyrolysis" is not correctly used when referring to this design. In the strictest sense, pyrolysis involves the

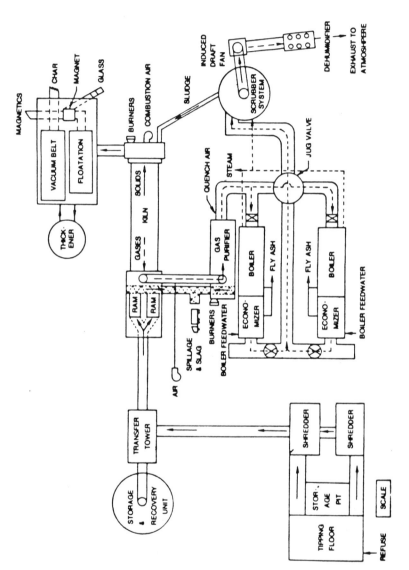

Figure 4-1. Schematic of the original Baltimore pyrolysis facility [16].

decomposition of a carbonaceous material *without* the use of oxygen, steam or carbon monoxide for direct heat transfer. The Baltimore system, however, supplies 40% of the stoichiometric air to the reactor to maintain a self-sustaining process and therefore alleviate heat addition via the process kiln burners [6]. The burners are needed during startup conditions to heat the reactor before waste feed is begun. Since it has become common practice to refer to the decomposition of material in an oxygen-deficient atmosphere as pyrolysis, this term is used throughout this discussion, although in this case "pyrolysis" actually means starved-air combustion.

The pyrolysis gases of approximately 650°C (1200°F) would then be combusted in the gas purifier. Heat was to be removed by two downstream boilers, producing a steam for sale to BG&E. The gas stream would then exit through a low-energy scrubber for pollution control before ultimate release to the atmosphere. Figure 4-2 shows several views of the plant as it existed during shakedown.

OPERATIONAL HISTORY

Shakedown and modifications to the Baltimore pyrolysis plant have, in effect, been an ongoing process since startup in January 1975. Major technical failures have plagued almost every unit process, making continuous operation impossible. With hardly any redundancy designed into the system (except for two shredder conveyors and shredders), specific unit shutdowns caused shutdown of the entire operation.

Some of the initial problems included: (1) difficulties with recovery of shredded wastes from the storage bin; (2) kiln refractory failure; (3) unstable kiln temperatures; (4) residue drag conveyor failure; (5) ram feeder jamming; (6) gas purifier slag taphole plugging; (7) induced air draft fan vibrations; and (8) high particle emissions. The plant was not able to meet the Maryland particulate standards of 0.03 gr/dscf @ 12% CO_2, which are stricter than federal standards of 0.08 gr/dscf @ 12% CO_2. It soon became apparent that technical, as opposed to operational, modifications were necessary. At this time, it was acknowledged that Monsanto was not meeting the previously agreed performance guarantees regarding air standards or plant operational capacities [7].

By November 1975, a supplemental agreement was made in which Monsanto agreed to contribute the $4 million in performance penalties and the EPA agreed to contribute an additional $1 million. This new financing was used to initiate a total of 96 modifications to the plant. These were to be done in two stages so that initial improvements could be evaluated and additional improvements added, if necessary [8].

Figure 4-2. Various views of the Baltimore–Monsanto pyrolysis facility: plant model (top); conveyors leading to storage silo (middle); and rotary kiln (bottom).

Unfortunately, further complications arose in April 1976 when the new improvements were being tested. These included: (1) erosion of gas purifier refractory; (2) extreme wear in the storage unit; (3) corrosion/erosion of the induced draft fan impeller and scrubber auxiliaries; and (4) inconsistent materials recovery operations [7]. Additional improvements were initiated from August to November 1976, including replacement of the kiln refractory and the residue drag conveyor.

By November 1976 operations were once again officially begun, and during the next three months the "plant operated at the highest level of throughput experienced during the demonstration" [7]. This was a throughput of 64,700 metric ton/yr (71,400 ton/yr), or about 245 metric ton/day (270 ton/day) at 6 day/wk at 85% operational capacity.

Problems still surfaced, however, such as gas purifier refractory failure and corrosion of the scrubber system. Significantly, it was decided to discontinue the material separation units while attempting to optimize the pyrolysis system. The residue building seems to have been too labor intensive (and therefore expensive), requiring six people during all shifts [9]. Further, it was decided that electrostatic precipitators would be needed to reduce stack emissions to an acceptable level. The average particulate emissions form the scrubbers in November 1976 were reported to be 0.255 gr/dscf @ 12% CO_2. This is roughly eight times greater than Maryland standards [10].

Despite the higher level of throughput, Monsanto decided to terminate its project involvement by February 1977. It also recommended that Baltimore end the recovery system attempt. A Monsanto official, D. L. Chapman, stated [11]:

> The major problems identified in January, 1976, have been solved. New problems, however, have surfaced and there exists the potential for further problems.

He noted that $6-$8 million of existing equipment could be used in converting the facility to an incinerator.

Mayor Schaefer, as insinuated before, was not pleased. Talk of taking Monsanto to court developed, although it appeared that Baltimore's hands were tied with the stipulation in the original contract [12]. Monsanto did provide the city with the updated plant design information, however, and paid the remaining $1.1 million of the $4 million not yet spent [11].

Although a consultant to the city also recommended a changeover to an incinerator, the city decided to continue to try and make the system work. A shutdown of a city incinerator since July 1975 due to its inability to meet air emission standards, combined with the paucity of city landfill volume, made ongoing waste disposal at the pyrolysis facility a must. Baltimore was producing roughly 1600 metric-ton/day (1800 ton/day) of MSW at 6 day/wk, and it could not afford to start anew [11].

With a commitment to make the plant work, the city began temporary measures such as patching the gas purifier refractory while deciding what final modifications and designs needed attention.

A board of advisors was formed of local officials knowledgeable in matters pertaining to the plant, including a president of a construction company, a Maryland Environmental Service official, a member of the Engineering Society of Baltimore, etc. [13]. Temporary measures were adopted while planning eventual modifications to keep the plant going.

Meanwhile, the EPA and the state Health Secretary exerted pressure on the system to come up with an air pollution control plan. Although the city had been warned several years before, the Schaefer administration was reluctant to invest such a large sum of money (estimates were up to $5 million) when the plant was only running intermittently [14]. Funding necessary to buy two electrostatic precipitators was thought to require a bond issue, which could not be voted on until November 1978 [15].

Although the facility was ordered closed in September 1977 because of the failure to have an adequate plan, it was allowed to operate five days later when state officials and the city came to an agreement. Plans for the air pollution controls were established, with interim measures of a high stack to be used until the electrostatic precipitators could be installed. A timely $3.1 million grant from the Economic Development Administration (EDA) helped to solve the financial problems facing the facility.

Figure 4-3 shows the most recent version of the pyrolysis facility. The four main modifications of the plant since Monsanto's departure are: (1) the replacement of the low-energy scrubbers with electrostatic precipitators; (2) the conversion of the gas purifier to a nonslagging operation; (3) the ultimate removal of the bin storage unit; and (4) the ultimate removal of the residue building [7]. Results from a test run in May 1979 were positive [1].

As of June 1979, the plant was running 6 days per week, processing 540 metric-ton/day (600 ton/day). Such tonnages are limited by the two electrostatic precipitators rated for 270 metric-ton/day (300 ton/day) each. Only minor problems with some refractory in the duct work and the reliability of the conveyance system have been encountered [1].

EVALUATION

In evaluating the Baltimore pyrolysis facility, it is obvious that technical inadequacies and some political and planning factors led to the problems experienced with the facility.

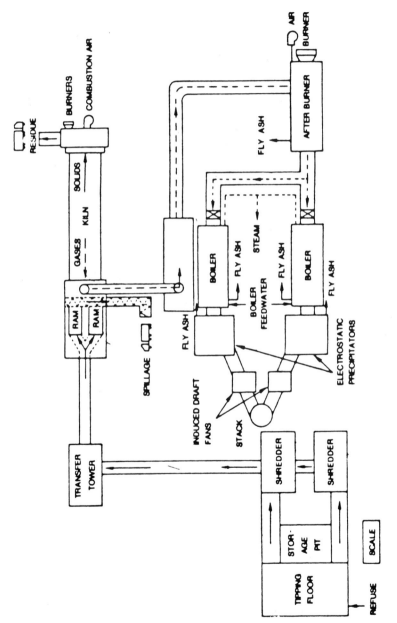

Figure 4-3. Modified schematic of the Baltimore resource recovery facility [7].

Technical Evaluation

First, the full-scale system did not perform in all respects like the 32 metric ton/day (35 ton/day) pilot plant, indicating some of the scaleup parameters were incorrect. Large scaleups of unknown processes are inherently difficult, and unless resources are invested in intermediate stages, chances will have to be taken with such large scaleups.

Second, and more important, is that the 32 metric ton/day (35 ton/day) facility differed from the full-scale version in 13 significant ways. Such deviations from the tested prototype add substantially to the operational risk of a system. Significantly, one of the main trouble spots of the Baltimore system was with one of these untested changes–a slagging gas purifier. The pilot-scale facility used a nonslagging purifier, and this was changed for the Baltimore facility as a final modification [7].

Third, Monsanto lacked information and experience with designing for MSW [7]. The heterogeneous and time-varying nature of solid waste require overdesign to provide for these variations. At Baltimore, "much of the material handling equipment had to be abandoned or required redesign" because of this [16].

Interestingly, Baltimore's problems in collecting small-particle-size particulates resembled the troubles that plagued the Nashville facility. Both systems used low-energy scrubbers, and both were eventually forced to change to electrostatic precipitators. In Baltimore's case, however, the formation of these unexpected smaller particulates has been attributed to the vaporization of oxidized metals and salts, which were then reoxidized to form a condensation aerosol. These would then be carried out with the pyrolytic gases [7]. These particulates were not present in the pilot facility [12], and it is suspected that this may have been caused by the MSW processed at the two facilities.

Planning and Political Considerations

Perhaps the most interesting factors discovered about the Baltimore demonstration facility were the personal interactions between those in the Public Works Department and associated agencies in Baltimore. On one side of the question were the staunch, proincinerator enthusiasts; the other side housed the less conservative, more resource recovery-oriented group. When initial problems occurred with the plant, it seemed that cooperation and/or patience was hard to find in some of the city officials. Two Baltimore pyrolysis employees were asked to resign–one a plant director and one a plant manager–between September 1976 and February 1977. This was during the critical modification period when Monsanto ultimately decided to leave the plant. It has been suggested that these actions were due in part to the continuing competition between the two groups [12].

The plant staffing, on the other hand, was even more complicated. Both Monsanto and city employees were on location, but the point of actual authority in different situations was not always clear. Effectiveness of the operating staff was consequently impaired. When Monsanto left the site, many of the city personnel had minimal experience with actually running the plant since this had been mainly Monsanto's responsibility. Confusion was compounded. Many other city staffing and organizational changes negatively affected the effective operation of the plant. Obviously, lines of authority and staffing should have been established in the original planning stages.

Although this planning inadequacy may not have mattered in the long run after modifications were complete and the system throughput rates had been decreased, it has been alleged that 900 metric-ton/day (1000 ton/day) of waste was not even available for processing when the plant was able to accept such quantities. Initial estimates of waste tonnages to be delivered to the site were consequently higher than actual tonnages available. It is therefore possible that modifications designed to increase the plant capacity to the initial design capacity of 900 metric ton/day (1000 ton/day) may not have been vigorously pursued.

Overall Evaluation

Baltimore was initially anxious to obtain the federal demonstration grant money [12], and it appears its vision was clouded when the word "demonstration" was written before the word "grant." As pointed out in a local newspaper, Monsanto may have "sold us a bill of goods," as Mayor Schaefer was quoted as saying [18]; but the city bought it. Dependence on an unknown technology will always have associated risks, as Baltimore has learned first hand.

It should be noted that Baltimore's redesigned pyrolysis facility may prove to be a technical and economic success. The rotary kiln unit has been shown to be an excellent primary combustion vessel, and it appears that Baltimore has been able to make the entire facility into a plant that both decreases waste disposal volumes for the city and produces a salable steam product that has a steady buyer. From an initial failure, Baltimore at last may have a success. However, a fair evaluation of the current version of the Baltimore pyrolysis plant is not possible until the system has been onstream for an extended period.

REFERENCES

1. May, E., Baltimore pyrolysis plant manager. Personal communication (1979).
2. Sussman, D. B. "Baltimore Demonstrates Gas Pyrolysis," EPA Publication Sw-75d.i, U.S. Government Printing Office, Washington, DC (1975).
3. Hale, S., Energy Development International, Washington, DC. Personal communication (1979).
4. *Baltimore Evening Sun* (November 10, 1971).
5. *Baltimore Sun* (January 7, 1977).
6. Helmstetter, A. J., and D. B. Sussman. "A Technical and Environmental Evaluation of the Baltimore Languard Demonstration," *Proc. 1978 Nat. Waste Processing Conf.* Chicago, IL (1978), pp. 447-55.
7. Systems Technology Corporation. *A Technical and Economic Evaluation of the Project in Baltimore, Maryland*, Vol II-IV, EPA Publication SW-175C (Springfield, VA: National Technical Information Service, 1979).
8. Sussman, D. B. "Baltimore Pyrolysis and Waste-Fired Steam Generator Emissions," *Waste Age* 7 (1976).
9. Ward, D., Baltimore pyrolysis plant operator. Personal communication (1979).
10. Hasebeck, J. L., and B. C. McCoy. *Source Emission Tests at the Baltimore Demonstration Pyrolysis Facility*. EPA Publication 600/7-78-232 (Springfield, VA: National Technical Information Service, 1978).
11. "Monsanto Also Gives Up," *Solid Waste Management* (3):54,79 (1977).
12. Zulver, E., pyrolysis plant director until September 1976. Personal communication (1979).
13. *Baltimore Evening Sun* (March 8, 1977).
14. *Baltimore Sun* (March 9, 1977).
15. *Baltimore Sun* (June 5, 1977).
16. Systems Technology Corporation. "A Technical, Economic, and Environmental Evaluation of the U.S. EPA Resource Recovery Demonstration Project in Baltimore, Maryland: Executive Summary," EPA Publication 719, U.S. Government Printing Office, Washington, DC (1978).
17. Sussman, D. B. Personal communication (1979).
18. *Baltimore Sunday Sun* (February 13, 1977).

CHAPTER 5

THE LOWELL RESOURCE RECOVERY PROJECT

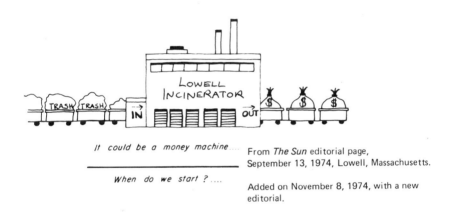

It could be a money machine.... From *The Sun* editorial page,
September 13, 1974, Lowell, Massachusetts.

When do we start ?....
Added on November 8, 1974, with a new
editorial.

The Lowell, Massachusetts incinerator residue processing scheme is one of the lesser known resource recovery projects initiated during the early 1970s because, in fact, it never got off the ground. Although EPA demonstration grant funding of $2.384 million was awarded to Lowell in 1972, and although engineering design of the recovery train was complete and construction bids had been let, the project was called off in 1975 essentially because of the failure to meet grant award requirements specifically set for the project. Of significance, however, is that the grant and subsequent project cancellation were not associated with the technical or economic viability of the recovery process itself.

Unlike the other initial recovery attempts, the Lowell project would have only recovered materials–not energy–from the solid waste. This was to be accomplished by processing incinerator residue from the Lowell incinerator and several other nearby incinerators through a series of materials separation

47

units. Design of the system was to be versatile enough to also accept air classifier heavies from a facility processing unincinerated wastes [1]. Revenues from the ferrous metal, aluminum, copper–zinc composite, sand and mixed-color glass products recovered from the noncombustible MSW in the incinerator residue were expected to offset annual debt service and O&M costs enough to obtain a profit from the operation. Inherent in the overall economics was the fact that the incinerator was already built and that any associated annual debt costs were not included in the recovery project economics. The unit processes themselves were tested using the residue from the Lowell incinerator in a pilot-scale recovery operation run by the U.S. Bureau of Mines (USBM). Resultant products from the pilot tests were used to help pinpoint markets, with market response and feedback used to ultimately design the order and/or degree of processing to be used in the full-scale Lowell demonstration plant.

Although markets, technical reliability and project costs appeared to point to a potential demonstration facility success, two major factors helped lead to the abandonment of the project before construction could be initiated. One factor known from the beginning of the project was that adequate air pollution control equipment for the Lowell incinerator had to be installed by 1975. This cost was not accounted for in overall cost estimates and was essentially disregarded by the city during the grant application period and early system planning. Another factor not apparent during initial planning stages was the significant decrease from original projections in the amount of incinerator residue available for processing. Residue from area incinerators—required to meet necessary design input volumes—became scarce when several incinerators were shut down due to their inability to meet new federal EPA emission standards. Project economics were consequently quite different from initial estimates because the purchase of necessary electrostatic precipitators substantially factored into capital costs, while reduced residue input would have led to decreased product volume recovery and revenues.

INITIAL FACILITY DESIGN AND ECONOMICS

Design of the Lowell facility was based on work done in conjunction with the USBM Edmondston, Maryland pilot plant by the facility designers, Raytheon Services Company (RSC). The pilot plant, completed by USBM in 1970, had a throughput capacity of 450 kg/hr (1000 lb/hr) of waste [1]. Design objectives included production of a highly consistent operation that could produce high-purity products. As mentioned previously, Raytheon was able to test the Lowell incinerator residue at the USBM pilot plant to aid in the design of the full-scale 225-metric ton/8-hr shift (250-ton/8-hr shift) demonstration facility and in the establishment of potential markets for the resultant material products. Such use of the facility was part of a cooperative agreement with USBM where RSC was allowed to use USBM design on a commercial scale as a result of paying the bureau an established fee and lending it staff [2]. In addition, tests were run with many associated full-scale units on a piecemeal basis utilizing the Lowell residue [1]. It is unfortunate that operational status of the designed system was never reached, as analysis of the system in light of the piecemeal, full-scale testing, design approach could have been enlightening.

Figure 5.1 depicts a simplified flow diagram of the as-designed Lowell-recovery project [1]. Incoming incinerator residue was to be directed through a trommel, where a glass-rich fraction (<0.64 cm; <0.25 in.) would separate from the remaining metal-rich fraction. Large metals would next be hand-picked from the trommel oversize (the metal-rich waste stream), followed by shredding and then primary magnetic processing of the oversize stream. To improve the possibilities of securing a market at steel mills, both the density and purity of the recovered ferrous would have been increased by an additional size reduction step, a washing process and, finally, a secondary magnetic separation.

The trommel undersize (the glass-rich fraction) was to have first been washed to remove ash and unburned material and then passed under a magnet to remove tramp ferrous metals (which are subsequently added to the trommel oversize waste portion) before further processing with the non-ferrous trommel oversize. These further steps were to include: (1) a density separation process (jig) where both heavy nonferrous metals (copper and zinc) and residual organics were to be concentrated separately; (2) a size reduction unit (grinder) where glass was to have been effectively crushed into small particles while at the same time only flattening, not decreasing the size of, the aluminum component of the waste; (3) a screening unit where separate glass- and aluminum-rich streams would have been produced; and (4) a series of glass processing steps where separation and cleanup processes (froth flotation, drying, etc.) were planned to produce marketable glass products.

It was expected that recovery of incinerator residue through the plant would decrease volumes of waste ultimately requiring disposal in the Lowell landfill to 30% of incoming volume and of dry weight. Future plans for the project included the reintroduction of the organic waste by products to the incinerator, ultimately reducing waste volumes needing disposal to 15% of original incinerator charge. Possible air pollution resulting from glass-dryer exhaust was to be alleviated through the use of a baghouse, and the 10^6 liters (270,000 gal) of wastewater expected to be produced each day was to be recycled as much as possible before treatment and subsequent discharge to the city sewerage system.

Initial system economic projections were developed for both 225-metric

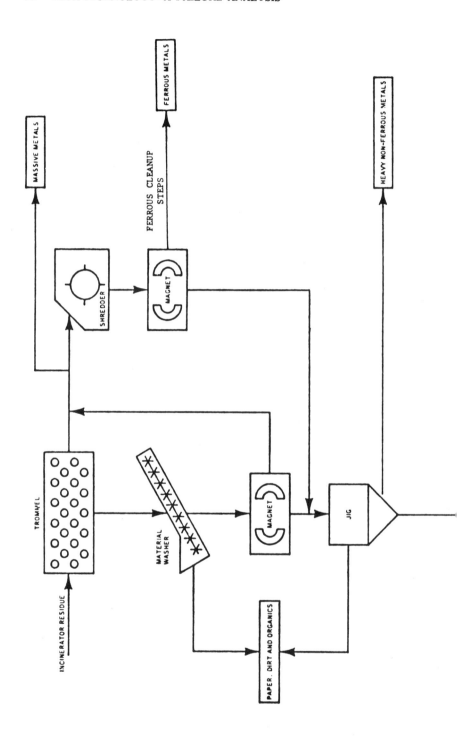

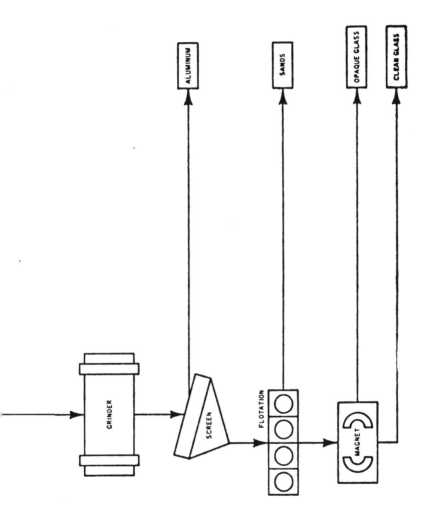

Figure 5-1. The Lowell incinerator residue processing operational schematic.

ton/day (250-ton/day) (one shift) and 680-metric ton/day (750-ton/day) (three shift) systems, resulting in a net disposal cost of $0.61/metric ton ($0.55/ton) and a credit of $5.57/metric ton ($5.05/ton), respectively, for the two systems [1]. It was assumed that only operational and maintenance costs increased for the 24 hr/day operation (Table 5-1). The assumption that equipment life is the same for both systems is debatable, however, perhaps necessitating an increase in capital costs for the three shift/day system as compared to those of the one shift/day operation to account for earlier equipment replacement. Additionally, revenues assumed for the system did not account for local market saturation for particular products. As discussed in the evaluation section of this chapter, it was found that markets for recovered material products of incinerator residue are difficult to locate.

PROJECT DEVELOPMENT

The city of Lowell was one of three cities considered for the incinerator residue recovery demonstration grant—the others being Chicago and Washington, DC. Although the decision-making process behind Lowell's selection as the grant recipient is not known, various sources have inferred that a main consideration was that the size of the Lowell operation provided the best assurance of a successful scaleup of the USBM pilot operation.

Table 5-1. Projected Operating Economics [1]

	One Shift/Day (250 tons of residue per 8 hr/day, 260 day/yr, 65,000 ton/yr processed)		Three Shifts/Day (750 tons of residue per 24 hr/day, 260 day/yr, 195,000 ton/yr processed)	
	Cost (Income) per year ($)	Cost (Profit) per ton input ($)	Cost (Income) per year ($)	Cost (Profit) per ton input ($)
Capital Cost[a]	424,000	6.50	424,000	2.20
Operation and Maintenance[b]	734,000	11.30	1,950,000	10.00
Total Cost	1,158,000	17.80	2,374,000	12.20
Revenue[c]	(1,122,000)	(17.25)	(3,364,000)	(17.25)
Net cost (Profit)	36,000	0.55	(990,000)	(5.05)

[a]Capital costs are amortized using an economic life span of 15 years and a 6% interest rate.
[b]Operating and maintenance costs are based on 11–12 persons/shift at an average salary of $10,000/yr.
[c]Revenue estimates are based on March 1975 material values.

Raytheon Services Company, already working with the city of Lowell in other projects [3], spearheaded the federal grant application and concurrently helped the city push for state funding assistance for the project. Massachusetts was developing a statewide resource recovery program at the time, and Raytheon had just completed the final version of the Massachusetts solid waste management report. Lowell's state application was not alone, however, as other assistance requests for proposed regional recovery plans were being submitted to the state during this period.

The incinerator residue recovery operation was originally presented to the state as part of a larger project for Lowell that also included incinerator upgrading [4]. Several Massachusetts Department of Natural Resources officials questioned the feasibility of funding the incinerator residue project from the beginning, however, citing some major problems [5].

Perhaps the most important criticism was that many felt the stricter air standards for refuse-fired incinerators, scheduled to be enforced in July 1975, would probably not be met by the current status of the Lowell incinerator. Therefore, possibilities of needing to upgrade the incinerator to meet emission standards or of closing it down altogether could easily render a new residue plant uneconomical or useless. The Massachusetts Department of Public Health felt the city of Lowell would not meet its department's December 31, 1972 deadline, which required an engineering report and implementation schedule for improvements in physical equipment and operations to meet air pollution control plans and regulations for the city's incinerator [6]. Evidently, the Public Health Department claimed the city had no engineers under contract as of June 1972 for meeting the deadline [6].

Recent discussions with regional state personnel indicate that Lowell's incinerator was in relatively poor shape at the time, having problems in controlling incinerator temperature, in getting complete combustion (burnout) of the waste,* and in unacceptable deterioration of the refractory lining. In addition, the air pollution control device used was a wet scrubber, which was not fully operational [4]. The fact that scrubbers are inadequate for complying with the stricter particulate standards was also recognized during this period,** indicating a need for more expensive air pollution control equipment, such as electrostatic precipitators.

Expressions of these misgivings regarding state funding assistance include the following excerpt from a July 1972 Massachusetts interdepartmental memorandum [7]:

> Aside from technical aspects to consider it seems premature to fund a residue recovery system before the plans for upgrading of incinerators have been approved.

---

*Bureau of Mines personnel have noted that Lowell residue samples processed at the USBM pilot plant showed various signs of insufficient burnout.
**See discussion in Chapter 2 regarding air emission control.

In another interdepartmental memorandum a week later, it was further stated that [8]:

> The proposed resource recovery process, in and of itself, is not objectionable. . . . However, this proposed process is only one part of a system of which the primary element is one or more incinerators. The proposed unit is inextricably tied to one or more incinerators. The proposal, in total, cannot be endorsed so long as it does not provide definite plans to upgrade the Lowell incinerator (or any other incinerator it is supplied by) to meet the requirements of state air and water pollution regulations.

Initial correspondence from the state to Lowell's city manager, James Sullivan, did not promise state monetary assistance to the city. Instead, general project support from the state's solid waste bureau was expressed, along with support for the city's application to EPA. Award of state assistance was mentioned as pending further review by various state agencies [9,10]. Copies of followup correspondence drafted to Sullivan from the state were found in the Massachusetts Department of Environmental Planning files. These draft letters reiterated previous expressions of general support for the project while stipulating that Lowell must be prepared to ensure that its incinerator meet air pollution regulations by a specified deadline of July 1, 1975 [11]. It has not been possible to establish whether versions of these draft letters were sent to the city; however, state personnel have confirmed that the city never got the state $0.6 million funding that various reports on the Lowell project mentioned [5]. It is likely, then, that stipulations regarding the meeting of air regulations were tied to the state award.

The official EPA demonstration grant award to Lowell in September 1972 included stipulations that resembled the concerns expressed earlier in the summer by state personnel. Specifically these qualifiers included [12]:

> The applicant must provide adequate assurance that the facility will receive a supply of residue consistent with the design capacity of the facility.

> The incinerator(s) from which residue is taken and processed by the planned facility should presently meet local, state, and federal regulations or guidelines for stack emissions. If not, the applicant must provide EPA with an adequate time-table and program for the incinerator(s) that will enable them to comply with those applicable regulations or guidelines.

> The applicant must provide assurances that it can provide the local share of the budgeted project costs and the total cost of any additional expenses required to complete the demonstration project.

It is not known if data regarding demonstration grant assurances were provided to EPA. An environmental review of the Lowell project was conducted under the auspices of the EPA, however, with the end result being the waiver of the need for preparing an Environmental Impact Statement (EIS) and the decision to award the grant [13]. A copy of this review, on file and available for public scrutiny at the EPA office on its completion in 1973, was buried in old files and could not be located in 1979.

Experimentation with, and design of, the project continued in 1973 and 1974, with Raytheon using USBM facilities and test marketing the resultant product streams. Quotes regarding market value were obtained for all the recovered materials except the sand product, with these used in estimating project economics (a value for sand was assumed). The marketing campaign was not a total success, however, as not all the project streams could be matched to definite or consistent markets. Apparently, RSC underestimated the difficulty of selling the products from an incinerator residue plant when it first began work on the demonstration project.

Interestingly, the city of Lowell either postponed or purposely decided not to initiate efforts needed for incinerator upgrading during this period. Part of this lack of action was undoubtedly associated with a form of political instability in the city, as the city manager's position changed hands in April 1974—in essence during the middle of the recovery facility planning process.

By the time the new City Manager, Paul Sheehy, took office in 1974, pressure to decide what to do about the city incinerator began to mount. A local newspaper issued a strongly worded editorial in September 1974 [14] criticizing Sheehy's administration for leaning towards abandonment of the incinerator project. Indicating that the incinerator "could be a money machine," as shown in the editorial cartoon on the first page of this chapter, the commentary mentioned:

> The Raytheon engineers have stated unequivocally that the plant can make money. . . . The cost of renovating the incinerator can be paid for ultimately, by the profit that the recycling operation is expected to make.

In addition, the editorial mentioned that the state of Massachusetts was thinking of routing the $600,000 grant originally earmarked for Lowell to a different regional recovery program [14].

The City Manager's lack of action regarding the incinerator recovery plant was criticized in stronger words in another editorial in November 1974 [15]. Reprinting the "money machine" cartoon and adding a subcaption of "When do we start?" the paper stated that the incinerator decision had been "on the back burner for an unconscionably long time now." Suggestions to continue with the original recovery plan—using a bond issue if necessary for incinerator improvements—were made.

Floating a bond issue did, in fact, appear to be the only way the city could meet the higher costs for continuing with the project. For instance, as consultant estimates for the necessary improvements came in, tipping fee quotes for Chelmsford—the largest user of the incinerator besides Lowell—were increased substantially from $6.60 to $14.30/metric ton ($6 to $13/ton) to make up for increasing costs at the Lowell incinerator. After meeting with Sheehy regarding the future disposal agreements, Chelmsford selectmen decided to investigate other disposal options and not be dependent on Lowell [16]. Other jurisdictions included in the initial waste flow estimates also began to reconsider their participation in the incinerator residue recovery project. These towns were also faced with meeting stricter air standards, with costs for the addition of more efficient air control devices being weighed against costs of such alternative means of waste disposal as landfilling.

It soon became apparent that improvement costs were indeed prohibitive. Estimates for incinerator and associated landfill improvements* ranged from $5 to $12 million [17]. Sheehy reportedly blamed Sullivan for neglecting the city's best interests by postponing badly needed incinerator improvements to hold the tax rate down during his incumbency [18]. Sullivan, however, indicated that while he held office he did not feel the incinerator improvements would be as expensive as the more recently quoted cost estimates [3].

A realistic appraisal of the residue recovery demonstration project's viability in light of both the expense of required incinerator upgrading and the decrease in waste flow available to the plant was finally determined as a result of March and April 1975 meetings attended by representatives of the federal EPA and the city. Major conclusions resulting from the meetings, as well as a request that the city of Lowell decline acceptance of the EPA demonstration grant, were expressed in an April 1975 letter to the Lowell City Manager from the Director of EPA's Resource Recovery Division [19]: Besides increasing cost problems, the EPA was

> . . . very concerned with the possibility of a decreasing supply of incinerator residue, the ability to meet air pollution standards, the useful life of the incinerator upgrading, schedule delays, and the ability of Lowell to meet EPA requirements of the original grant.

The 1975 daily available incinerator residue was established as only 136 metric ton/day (150 ton/day) as opposed to 225 metric ton/shift (250 ton/shift) design. As far as the Director knew, there were

> . . . no written agreements or statements of interest from even those three communities which we are now including in our residue estimates.

The total disposal costs for Lowell were approaching the highest in the county. Besides paying the costs for incineration (estimated as $32/metric ton ($29/ton) including improvement costs and based on 180 metric ton/day (200 ton/day) over 15 years), the city would also have to pay residue plant costs projected as $4.60/metric ton ($4.20/ton) for a 165-metric ton/day (150-ton/day) throughput.

The memorandum went on to state that the EPA would not provide Lowell with any additional funds above the original grant award of $2,384,000.

> There is no guarantee that the incinerator will meet air pollution standards after completion of the upgrading. . . . Both the State of Massachusetts and EPA air pollution regional authorities have recommended strongly that Lowell

---

*Excessive leachate production necessitated the installation of liner at the landfill.

not try to continue operation of its incinerator by upgrading. Consequently, both are unlikely to be willing to allow variances should the upgraded incinerator still not meet air standards.

All issues in the original grant award letter had not been addressed. For example, local share of costs necessary to meet the "additional expenses required to complete the demonstration project" were estimated as necessitating a $6.75 million bond issue.

> The magnitude of the combined effect of these problems is great. We (EPA) feel it difficult to encourage the City of Lowell to continue with this project. The project costs would force Lowell to pay one of the highest disposal costs in the country. This is obviously not the intent of the EPA resource recovery program.

As a result of this letter and of the previous meetings regarding the project, the Lowell City Council voted on May 20, 1975 to "expend no further funds" on incinerator improvements and to withdraw from the demonstration grant [20]. The council cited the "inflationary costs of the project."

EVALUATION

This account of the attempt to implement a resource recovery facility in Lowell, Massachusetts shows a combination of factors leading to the final "no-go" decision for the project. The most telling of these factors—the unfavorable, bottom-line project economics—included capital outlays specific to the city of Lowell, however, such as the expensive incinerator upgrading requirements. Whether costs would remain prohibitive for a similar facility in another city having more favorable initial conditions and having markets for the recovered materials has yet to be determined. Significantly, however, it must be remembered that the technical feasibility of the full-scale residue recovery project still has not been demonstrated.

To benefit from this chapter of resource recovery history, the planner should pose and then answer as best as possible three questions related to Lowell's unsuccessful attempt at resource recovery.

1. What could have helped get the residue plant "off the ground" for the city of Lowell?

Perhaps the most obvious answers to this, in light of the 20/20 vision of hindsight, are both the proper maintenance of the city incinerator throughout its life and the upgrading of the incinerator during early system planning (as soon as the federal grant award was official). Such a maintenance program ultimately might have decreased the needs for incinerator improvements and thus reduced the project cost to Lowell. In addition, an early start on the upgrading would have undoubtedly boasted lower costs than those quoted later because the combined effects of inflation and years of below-par incinerator operation would have been minimized. Although total upgrading costs might still have been prohibitive for the project had these two steps been taken, at least the realization of the costs that would have had to be borne by the city would have occurred at a much earlier time. Perhaps alternative means of funding could then have been pursued.

The presence of an individual or group of individuals controlling the project—having the power to analyze, set priorities, then initiate necessary project inputs from the city—would have also increased chances of project implementation. Steps for meeting incinerator improvements could have been established and followed, with realistic assessments of cost requirements to the city determined. Thus, this project management board could have assessed the potential needs for the issuance of general obligation bonds as well as minimized any detrimental effects to the project implementation as a result of the change of hands in the City Manager's office during the middle of the project planning period.

2. How could such an unfruitful period of planning have been avoided, or at least shortened, for Lowell?

Most of the above discussion is relevant here as well. The early scheduling of incinerator improvements could have quickly pointed to a no-go decision for the city, for instance, had costs been found to be prohibitive. Further, a project management team could have clearly projected necessary inputs required of the city by early project planning stages, therefore providing a data base which the city could use to analyze whether to continue with the project. Besides the use of these early project planning tools to shorten the unproductive project planning period, however, a realistic and pragmatic view should be given to a potential resource recovery project. Lowell, for instance, was in a position where its municipal waste disposal techniques were inadequate. The incinerator was in need of repair as well as under state order to upgrade its emission control abilities by July 1975. The landfill used for disposing of the incinerator residue was also in poor condition and was experiencing major leaching problems. Although the planned recovery plant would have decreased the amount of incinerator residue requiring disposal, it could not approach the initial, much higher volume waste reduction capabilities of the incinerator. In terms of waste disposal, the funding for the residue recovery plant by the city should probably have been placed as second priority compared to upgrading the incinerator.

The costs to construct the recovery plant seemed at first to be covered by the federal and pending state funds, and this may have made the residue plant appear to be a feasible undertaking. Nonetheless grant stipulations that the city incinerator meet environmental standards and that the city supply any extra associated project costs should have made it obvious that the city would be required to provide upfront costs to upgrade the incinerator before the implementation of the residue plant. The projected high project revenues that could have been used to help defray the incinerator improvement costs were dependent on a three shift per day system—a system requiring a large regional incinerator waste network that was never established for the area. The smaller, one-shift 225-metric ton/day (250-ton/day) network established for Lowell could not come close to the benefits derived by the economics of scale evidenced in the 750 ton/day system, however. In essence, then, only a small net revenue, if any, could be expected from the size system to be demonstrated at Lowell.* As the communities included in the network began to pull out from the system, projected throughput tonnages to the system were decreased, as were the total projected revenues from recovered materials.

3. Does incinerator residue recovery have the potential to be a feasible undertaking in another location?

The answer to this question is twofold. First, as mentioned previously, the full-scale application of this processing train has not yet been demonstrated. Use of this system would therefore include some risk, and the potential for spending large amounts of time and money in scaledown procedures exists for any application.

Second, every area has site-specific characteristics that contribute to the overall applicability of a system to that area. Lowell's need for costly incinerator improvements combined with the decreasing area incinerator residue tonnages made a residue recovery system unfeasible for the area.

Many participants in this project have expressed regret that the time and money expended produced nothing useful—the technology is yet to be proven and Lowell's solid waste disposal problems were not resolved. The failure of the project can be traced to poor planning and overenthusiasm. Markets were not adequately analyzed, and factors such as the condition of the incinerator and the availability of solid waste were not factored into the decision-making.

Unfortunately, the "money machine" was never cranked up.

REFERENCES

1. Arella, D. G., and Y. N. Garbe. "Mineral Recovery From the Non Combustible Fraction of Municipal Solid Waste," EPA Publication SW-82 d.l, U.S. Government Printing Office, Washington, DC (1975).

2. Easterbrook, G. E. "A Natural Environment for Resource Recovery," *Waste Age* 8 (1978).

---

*Although the figure cited by the EPA in 1975 mentioned a net cost of $0.55/ton residue in a one-shift plant, these calculations incorporated recently projected increases in plant cost due to inflation since the project's conception [1]. Earlier net cost projections may have claimed a net revenue per ton for the 250 ton/day system.

3. Sullivan, J. L., Lowell City Manager until April 1, 1974. Personal communication (1979).
4. Dainall, K., Massachusetts Department of Environmental Quality Engineering, Merrimac Regional District. Personal communication (1979).
5. Cousins, A. E., Bureau of Solid Waste Disposal, Massachusetts Department of Environmental Management. Personal communication (1979).
6. Memorandum to A. Cousins from R. Power, Department of Public Health (July 13, 1972).
7. Memorandum to A. W. Brownell, Chairman of Water Resources Commission from A. E. Cousins, Director of Office of Environmental Protection (July 6, 1972).
8. Memorandum to A. W. Brownell, Chairman of Water Resources Commission from A. E. Cousins, Director of Office of Environmental Protection (July 14, 1972).
9. Letter to J. Sullivan, Lowell City Manager, from Bruce Campbell, Massachusetts Commission of Public Works (July 14, 1972.
10. Letter to J. Sullivan, Lowell City Manager from R. Bennett, Assistant Secretary for Administrator, Massachusetts Office of Planning and Budget (July 14, 1972).
11. Draft of letter to J. Sullivan, Lowell City Manager, from R. Bennett, Assistant Secretary for Administration, Office of Planning and Program Coordination (July 1972).
12. Correspondence to J. Sullivan, Lowell City Manager, from Samual Hale, Deputy Assistant Administrator for Southwest Management Programs (1972).
13. Correspondence received by Massachusetts Water Resources Commission, Office of Environmental Protection, from Hale, Deputy Assistant Administrator (January 4, 1973).
14. The Lowell Sun (September 13, 1974).
15. The Lowell Sun (November 8, 1974).
16. The Lowell Sun (September 11, 1974).
17. "Resource Recovery Systems . . . A Review," NCRR Bull. 5(2) (Spring 1975).
18. The Lowell Sun (March 19, 1975).
19. Letter to Paul Sheehy, Lowell City Manager, from Nicholas Humber, Director Resource Recovery Division (April 29, 1975).
20. Letter to Humber, Director of EPA Resource Recovery Division, Office of Southwest Management, from P. Sheehy, Lowell City Manager (June 19, 1975).

CHAPTER 6

THE SAN DIEGO FLASH PYROLYSIS PROJECT

Ashes to ashes
Tar to tar
Too bad we can't get pyro oil
To put in a jar.

David Sussman*

This version of the "ashes to ashes" rhyme seems to perfectly sum up the San Diego flash pyrolysis demonstration project. With respect to the main objective of the demonstration, that of investigating a full-scale application of a pyrolysis technique that produces a marketable fuel oil from MSW, the project was a success. The pyrolysis technique was not, however. Although roughly 15,300 liters (4050 gal) of pyro oil were produced, the "fuel" was unmarketable and unusable [1]. Occidental Research Corporation (ORC), designer and largest funder of the $14.4 million project, closed the facility in July 1978. It has been estimated that ORC may have invested up to $9 million in the project by the time the facility was closed, however, increasing project costs to roughly $15 million.**

The facility incorporated a thorough preprocessing of wastes before pyrolysis, separating out glass, aluminum and ferrous materials. Although these seemed to operate well, the project evaluators could not economically

---

*This rhyme was found on an aerial shot of the San Diego facility hanging in the EPA Solid Waste Office and is attributed to David Sussman who was a member of the EPA Office of Solid Waste Management team involved in the San Diego Project.
**Contributions to the project were roughly $2 million from San Diego County, $4.2 million from the EPA, and $8.2 million from ORC. A detailed account of expenditures in an April 1979 report gave a $14.47 million figure [1].

recommend further adoption of the glass recycling system. They described the recovered aluminum product as "less than desirable" and spoke of difficulties encountered in recovering a high-purity ferrous product.

What, then, can be concluded from the San Diego demonstration? Is the pyrolysis of MSW to form a fuel oil an uneconomical and technically unfeasible prospect? Presently it seems so. E. R. Holderness, Occidental's project manager for the facility, agrees with this prognosis [3].

BACKGROUND INFORMATION

Unlike many areas, the county of San Diego seems to have always enjoyed the abundance of relatively inexpensive landfill for disposal of its wastes. With landfill costs of $1.70/ton when the contract between the county and ORC was initiated [4], the municipality paid only $5–$7/ton for land disposal in 1979 in landfills of projected 20-year lives [5]. Why, then, did the county become interested in housing and partially funding a demonstration resource recovery facility? It appears that the county wanted to develop alternative energy sources for use during times of conventional energy shortages [4]. Since the San Diego area is highly dependent on oil for heating and transportation, the Occidental flash pyrolysis system, designed to produce a marketable fuel oil, was seen as a logical project to support.

Financially, San Diego County could readily afford to invest in the pyrolysis facility. Although it owned and operated its own landfills, only a small percentage of tax revenues was spent on solid waste management because most of the solid waste collection was conducted on a user-charge basis by annually licensed private firms. A large tax base, coupled with a large, unutilized debt capacity, added to the county's ability to ultimately spend $2 million on the energy alternative demonstration facility [4].

ORC (formerly Garrett Research and Development Company) began research and development of the flash pyrolysis process in 1968, and by the following year was experimenting with a 1.4 kg/hr (3 lb/hr) bench-scale unit that converted organic materials to liquid fuel. Based on an outgrowth of research in coal liquefaction, the unit tested such wastes as bark, rice hulls, sewage sludge, waste rubber, animal manure and finely shredded MSW [6,7]. By March of 1971, Occidental had constructed a 3.6 metric-ton/day (4 ton/day) unit in Vancouver, Washington to facilitate both the establishment of fuel product characteristics and the investigation of critical process variables and materials handling problems [6,7]. The pilot facility was subsequently moved to ORC's La Verne, California location. The La Verne pilot plant's capacity was reported to be as large as 7 ton/day [8].

From the early stages of development, ORC had also included glass and metals recovery in its pyrolysis plans. A major consideration in the design of such separation processes was to produce a high-purity product, thus increasing ultimate product marketability.

Following the demonstration grant award to San Diego County, a turnkey contract was initiated between ORC and the county. The contract held Occidental responsible for design, construction, startup and the development of the facility into a fully operational plant before turning it over to its ultimate owner—San Diego County. Occidental subcontracted to the Ehrhart Division of Procon, Inc., a subsidiary of Universal Oil Products Inc., for assistance in parts of engineering design, procurement and construction [9].

Difficulties occurred in finding a site for the project, however, as communities in the county seemed reluctant to house the facility. Of 15 communities approached, the city of El Cahon finally leased the county a 5.3-acre industrial plot [10]. The city initially charged the county a nominal $1/yr fee during the demonstration period, although it increased the lease to reflect market prices of $32,000/yr in the spring of 1978. The county was subsequently able to convince the City in September 1978 to decrease the lease fee to $16,000/yr [11].

Unfortunately, inflationary factors resulting from the delay in finding a suitable location and in securing various permits helped increase original estimates of project capital costs from $6,344,000 to $11,300,000 before construction was begun in February 1976. Some other factors affecting such a major cost increase included the need for additional odor control equipment, the addition of aluminum recovery processing, site-induced design limitations* and previously unaccounted for landscaping cost [15].

The delay in finding a site also put the implementation phases of the demonstration program behind schedule [6]. Originally slated for August 1975, the construction phase began February 1976. The final phase of the project, a one-year testing and evaluation period scheduled to begin September 1976, actually began in August 1977 when the third-party evaluation was started. Although postponed by previous construction delays, the evaluation phase was also behind schedule due to the need for an unscheduled seven-month shakedown period.

---

*One of the most interesting site-induced limitations is a meteorological condition that prevails at certain times, creating air stagnation and NO_x buildup. When this occurs, NO_x stack-emitting industries in the area are forced to close until the condition subsides. Since the demonstration facility could be included in such a measure, the county bought three air-monitoring units to monitor for such conditions in the plant vicinity [12].

A project schedule of October 1974 [6] provided only a one-month period between construction and the start of the evaluation phase. Unless the evaluation phase was conceived as a shakedown period, the original project scheduling was overly optimistic, as all resource recovery facilities seem to need a period of shakedown before trying to meet design rates and throughputs.

FACILITY EXPECTATIONS

The San Diego County pyrolysis demonstration project was designed to convert 180 metric ton/day (200 ton/day) of MSW into marketable fuel oil, aluminum, glass and ferrous metal. Table 6.1 provides a breakdown of the projected quantity of products, as well as unmarketable by-products, produced in a design day. As described later, much of the solid and gaseous by-products were to be recirculated for use in plant processing.

The pyrolytic fuel produced at the Occidental facility was to enter the market as a substitute for No. 6 residual fuel oil in oil-fired, steam-generating boiler furnaces. Successful combustion tests, made by a Combustion Engineering, Inc. laboratory using 50 and 25 volumetric proportions of pilot plant pyrofuel with a No. 6 fuel oil, gave credence to the proposed market [6]. Arrangements to purchase and burn the fuel from the 180 metric-ton/day (200 ton/day) plant (cofiring it with residual oil) in a San Diego Gas & Electric Company 445,000 kg/hr (980,000 lb/hr), 14,800 kPa (2150 psig) 540°C (1000°F) superheated steam boiler were initiated [13], with burning scheduled to begin when adequate storage and fuel handling equipment were installed at the utility. All associated capital expenditures incurred by the Gas & Electric Company were to be based on equivalent fuel values determined by the utility [6]. Resultant combustion

Table 6-1. Projected Outputs from the Flash Pyrolysis Facility [8]

Product	wt %	Amount/Day
Fuel Oil	25.6	235 barrels[a]
Aluminum	0.4	0.7 metric-tons (0.8 tons)
Glass Cullet	5.3	9.5 metric-tons (10.6 tons)
Ferrous Metal	6.7	12.1 metric-tons (13.3 tons)
Flue Gas	24.5	
Solids (including char)	16.6	30.2 metric-tons (33.2 tons)
Wastewater	20.9	38.0 metric-tons (41.8 tons)

[a]The literature varies in projections of fuel oil production. Discounting the 14 wt % water content that was to be included in the product, only 194 barrels would be produced (adapted from Preston [8]).

characteristics of various cofiring ratios and associated emissions were then to be established [13].

As shown in Table 6.2, a comparison between the characteristics of the pyrofuel produced in the pilot tests and those of a No. 6 fuel oil show the volumetric heating value of pyrofuel to be 77% of the conventional fuel. A 14% (by weight) water content, retained in the fuel to decrease its high viscosity and consequently improve its handling properties, would decrease the volumetric heating value to roughly 64% of a No. 6 oil, however. The viscous pyro-oil requires higher temperatures than a No. 6 for ease in pumping, storage and atomization, although tests showed that maintaining the fuel above 71°C (160°F) for extended periods could irreversibly increase fuel viscosity [8].

The pyrofuel also has a low sulfur content, a 60% solubility in water and a corrosive effect on mild steel due to its somewhat acidic nature. Most of the mentioned fuel characteristics are felt to be linked to the high oxygen content of the product [8].

Expectations for the other recovered products include 75% recovery of incoming glass in a 99.5% pure, mixed-color form; 60% recovery of incoming aluminum in an 85–95% purity range [14]; and 95% recovery of incoming ferrous materials of a 95% purity [6].

The 180 metric-ton/day (200 ton/day) demonstration facility was never touted as being a cost-effective treatment of municipal waste. A 1975 cost projection for the plant without aluminum processing or recovery was

Table 6-2. Comparison of No. 6 Fuel Oil and Pyrofuel [8]

Property	No. 6 Fuel Oil	Pyrofuel from the Pilot Plant	
Carbon (wt %)	85.7	57.0	
Hydrogen (wt %)	10.5	7.7	
Sulfur (wt %)	0.7-3.5	0.2	
Chlorine (wt %)	–	0.3	
Nitrogen (wt %)	2.0	1.1	These values do not include 14 wt % water.
Oxygen (wt %)		33.2	
Ash (wt %)	0.05	0.5	
Specific Gravity	0.98	1.30	
Volumetric Heating Value (Btu/gal)	148,840	114,900	
Heating Value (Btu/lb)	18,200	10,600	
Density (lb/gal)	8.18	10.85	
Pour Point (°F)	65-85	90	
Flash Point (°F)	150	133	These values do include 14 wt % water content.
Viscosity @ 190°F (Ssu)	340	1150	
Pump Temperature (°F)	115	160	
Atomization (°F)	220	240	

$14.91/metric ton ($13.42/ton) of incoming waste [6]. It was thought that larger scale versions of the flash pyrolysis facility would be necessary before economies of scale would create an economically competitive facility. A 1976 cost estimation for similar facilities of 1000 and 2000 ton/day capacities that also employ aluminum recovery assigned net costs of $12.77 and $6.58/metric ton ($11.49 and $5.93/ton), respectively [8]. Later cost projections in a 1978 report estimated 1000 and 2000 ton/day facilities as having net costs of $15.57 and $8.51/metric ton ($14.12 and $7.72/ton), respectively [7]. A major component of the third-party evaluation report on the 180 metric-ton/day (200 ton/day) system was to be a realistic cost estimation of both 1000 and 2000 ton/day versions of the El Cahon facility, incorporating actual plant results, operation costs and market acceptance of the recovered products [1].

FACILITY DESCRIPTION:

During periods of full operation, the demonstration facility was designed to accept MSW 8 hours a day for 6 days a week [1]. Operation of the primary shredder, the electromagnet and the stacking conveyor were to occur only during the 8-hour day, while all other unit processes were scheduled for continuous 24 hours a day, 6 days a week operation.

A flow diagram of the Occidental facility, as seen in Figure 6.1, depicts the complex and highly interrelated design of the system. All major unit operations found in the preprocessing and the pyrolysis systems are represented, although the glass and aluminum recovery systems have been greatly simplified. The facility is pictured in Figures 6.2 and 6.3.

Once the refuse is dumped on the receiving floor, a front-end loader feeds the MSW to a primary shredder, a 32 metric-ton/hr (35.5 ton/hr) capacity, 750-kW (1000 hp) horizontal hammermill of manganese-steel hammers. The waste stream, now of nominal 7.5-cm (3 in.) particle size, passes under an electromagnet to remove 95% of the incoming ferrous materials and continues to a rotating stacking conveyor.

Unique to this facility, the stacking conveyor directs one-third of the shredded waste to another conveyor feeding the rest of the facility and evenly distributes the remaining two-thirds waste on a concrete floor below it by revolving in a 360° fashion while discharging the shredded feed. Waste thus deposited is fed to the conveyor by a front-end loader during the remaining two shifts. The waste is then dropped into a doffing roll bin, new to the MSW processing field, which acts as a temporary storage bin and as a metering device for the feeding of the air classifier. The spiked delumper rolls in the bin provide an even feed as the waste is conveyed

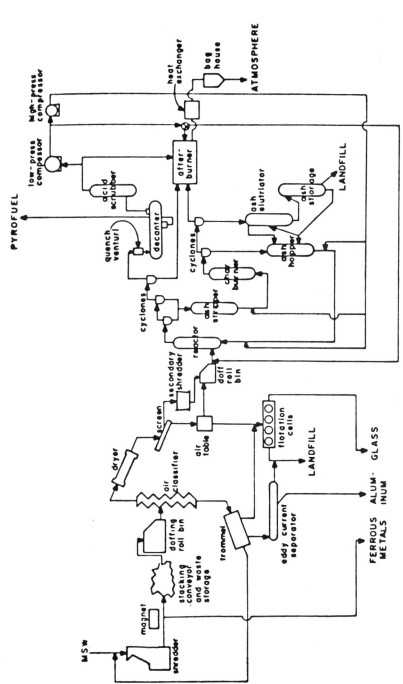

Figure 6-1. The San Diego flash pyrolysis operational schematic.

Figure 6-2. Artist's sketch of the completed San Diego flash pyrolysis facility as it would have looked.

between them and the floor drag chain [14]. From the doffing roll bin on, all waste processing equipment is located outdoors.

The 13-meter (43 ft)-high, 8 metric ton/hr (9 ton/hr) capacity, 10-stage, zig-zag air classifier, designed by Occidental, separates the light, pyrolysis-directed waste components from the heavy, aluminum and glass-rich fraction of the waste—roughly a 70/30 split [1]. Also unique to the ORC facility is the recirculation of cyclone air in the classifier [14].

The heavy stream passed through a 3-meter (10 ft)-long, 12 rpm, 1.42 meter (4.67 ft)-diameter trommel of 3.8 cm (1.5 in.) square and 10 cm (4 in.) round holes, separating the heavies into glass- and aluminum-rich streams, respectively. Particles larger than 10 cm (4 in.) are returned to the primary shredder. The fundamental process of eddy-current separation and froth flotation (a proprietary reagent is used) are employed in the production of the high-quality products [9,14].

More processing of the highly organic, light fraction is needed to produce the relatively dry, small-particle-sized, strictly organic feed desired for optimal pyrolysis performance. The waste moves through a rotary kiln dryer, decreasing moisture content to roughly 3.5%, and enters a vibrating screen in an inert gas atmosphere, where ultimate removal of 85–95% of incoming inorganics is completed [1]. Waste particles greater than 0.83 mm (20 mesh) are directed to the secondary shredder, a horizontal plate, inert gas atmosphere attrition mill for reduction to a nominal particle size

Figure 6-3. The pyrolysis plant under construction. Oil fractionation tank is in middle of the picture, with the oil storage tanks behind the main structure (courtesy of F. Macalister).

less than 1.17 mm (14 mesh). The screen undersize is processed in an air table, producing a light, organic fraction to join the secondary shredded material in the secondary doffing roll bin, a heavy, inorganic fraction to supplement the glass recovery feed; and a medium-weight reject requiring disposal. The waste in the inert gas pressurized doffing roll bin is now ready for processing in the pyrolysis system.

As shown in Figure 6.1, the pyrolysis system is composed of a complex network of recirculating loops that incorporate the char and pyrolytic gas produced during the flash pyrolysis process. Only during startup is an outside supply of spent fluid cracking catalyst and nitrogen gas used.

The flash pyrolysis cycle begins when processed waste and low-pressure recycle gas enter a 6.8-meter (22.2 ft), refractory-lined vertical reactor vessel and are pyrolyzed under 510°C (950°F) 83 kPa (12 psig) conditions for 0.3 seconds [1]. Hot (730–760°C; 1350–1400°F) recirculated ash, fluidized with high-pressure recycle gas, is used as the indirect heating medium in proportions of roughly five pounds of solids per pound of waste fuel [15].

Three series of cyclones provide solid–gas separation immediately following the flash pyrolysis reaction. The pyrolytic vapors are quickly transported to a quench venturi, where intimate mixing with a No. 2 quench oil terminates further chemical breakdown (cracking), condensing much of the vapor into a pyrofuel mixture. The pyrofuel and quench oil fractions are subsequently separated via their different specific gravities in a decanter. Noncondensible pyrolytic vapors are cleaned in an acid scrubber at 54°C (130°F), 41 kPa (6 psig), and then either compressed in low-(205 kPa; 30 psig) and high-(308 kPa; 45 psig) pressure centrifugal compressors for use as an inert transport medium or combusted in the afterburner for process heat purposes. The solids from the third cyclone are also combusted in the afterburner.*

The solids (char and catalyst) separated in the first two cyclones are deposited via free-hanging dip legs into the ash stripper, a vertical, refractory-lined vessel operating with identical design temperature and pressure as the pyrolysis reactor. Char collected in this vessel enters a cold recirculation loop and is pneumatically transported by high-pressure recycle gas to the char burner, where any unburned organics are combusted. This vessel, fed with combustion air heated in the afterburner, has design pressures of 12 psig and temperatures of 730°C (1350°F). The gas-solid stream produced is separated in two more cyclones, the final gas product combusted in the afterburner [1].

Solids from both char burner cyclones feed the ash hopper and ash elutriator, respectively, via free-hanging dip legs. Some of the ash collected in the refractory-lined elutriator may feed the ash hopper, although most is cooled with water and ultimately transported to the ash storage vessel. Operation of the refractory-lined ash hopper completes combustion of remaining char, if any, and heats the ash to 730°C (1350°F) under pressures of 68 kPa (10 psig). The ash, along with high-pressure recycle gas, is pneumatically conveyed through a refractory-lined, hot ash circulation loop to the pyrolysis reactor for completion of the char recirculation cycle. Solids stored in the ash storage vessel are used to supplement ash quantities in the hopper or elutriator, with excess quantities periodically discharged for disposal [1].

As previously mentioned, the pyrolytic char product is combusted and used as a heat transfer medium in the flash pyrolysis process. The compressed pyrolytic gas is used as an inert transport medium; as a fuel in the refractory-lined afterburner to provide heat to such processes as the rotary kiln dryer and char burner; and to oxidize at 650°C (1200°F) for 0.5 seconds any combustible or odorous pollutants from plant processes, including the rotary dryer and char burner gas streams [8].

The afterburner, then, acts as an antipollution device as well as a

---

*There is some uncertainty about these pressures. The evaluation report [1] gives values for pressure of the gas exiting from the high-pressure compressor as both 45 psig and 65 psig. The former has been chosen for use in this discussion.

process heat supplier. Provisions to burn No. 2 fuel oil in the unit are available, although the afterburner is designed to first depend on recycle gas and then pyro-oil for fuel. Four baghouses are used for air pollution control, one at the primary shredder, one for the solid processing units of the screen and air table, and two for the afterburner air stream [1].

Unusable solid by-products, such as excess ash and the aluminum and glass plant rejects, are landfilled. The liquid effluent from the glass plant is treated and then recycled, with periodic additions of makeup water [13]. Water products from the pyrolysis process, having high chemical oxygen demand potentials [8], are ultimately disposed into the sewerage system. Although originally no provisions were made for water pollution control, pretreatment may be required before disposal into sewers.

SHAKEDOWN AND OPERATIONAL EXPERIENCES

Initial shakedown activities, conducted January through April of 1977, focused on the front-end, preprocessing units of the primary shredder through the air classifier and trommel. Startup of the glass and aluminum recovery systems was then initiated, followed by the start of the remaining waste feed preparation units [13]. Pyrolysis runs began in August and continued through mid-March of 1978, after which plant staffing was roughly cut in half [16]. During the pyrolysis operation, five unsuccessful attempts were made to pass an EPA acceptance test of 72 hours of consecutive pyrolysis operation. Additional EPA funding of between $500,000 and $360,000 was expected if the runs had been successful [17]. Various parts of the nonpyrolysis systems continued in operation until staffing of the demonstration facility was completely discontinued the end of July 1978 [1].

Nonpyrolysis System

The operations of the processing units up to the trommel, excepting the electromagnet, appear to have had normal shakedown experiences. Relatively minor alterations or adjustments enabled the units to reach near-design processing expectations.

An increase of the grate spacing in the primary shredder from 7.5 X 15 cm (3 X 6 in.) to 15 X 23 cm (6 X 9 in.) allowed a closer attainment of the 32 metric-ton/hr (35 ton/hr) design throughout without significantly degrading the desired size distribution of the shredded product. Although the concept behind the rotating stacking conveyor was successful, alterations to give the 7 meter (56 ft) arm a bidirectional rotation enabled quicker positioning of the arm over the conveyor feeding the doffing roll bin [1].

Initial leaking of rain into the doffing roll bin (since it is located outdoors) was alleviated by sealing off all openings. Bending the tips of the conveyor teeth at 0.3-meter (1 ft) intervals decreased the initially high shearing action exerted by the teeth on the waste (which had contributed to higher loads on the drag chain of the conveyor) and consequently increased the feeding abilities of the bin [1].

As the air classifier is a unit process whose performance is affected by many factors, considerable data were obtained during shakedown for use as an aide in achieving desired operational conditions. Parameters that were varied included doffing roll bin chain setting and degree of damper opening; resultant data areas included waste feedrate, pressure drop and percentage light/heavy split. Corresponding underflow (heavy) component breakdowns were established with respect to aluminum, magnetic material, glass, trommel oversize and waste content.

Initial experiences with the trommel gave a higher glass content in the aluminum stream than desired. Addition of a small baffle plate at the point of size change in trommel screen openings helped to increase glass residence time in the trommel, allowing greater glass capture in the glass-rich stream.

The alterations and adjustments instigated in the one-stage ferrous recovery system, however, had limited success in decreasing the high organic content of the final ferrous product. Improvement steps initiated included addition of a suction bypass line in the ferrous discharge chute for collection of organics, installation of a vertical, adjustable baffle 0.5 meter (1.5 ft) from the vacuum line to promote movement of organics back to the suction line, addition of yet another baffle, and optimization of conveyor belt speed under the magnet. These modifications were able to remove many of the organics greater than 5 cm (2 in.) from the ferrous metal product.

As inferred at the beginning of this chapter, the shakedown and operational experiences of the aluminum and glass recovery systems were not encouraging. The "less than desirable" quality of the aluminum product significantly affected revenue-generating abilities of the system as potential buyers quoted roughly one-half originally expected values for the aluminum product [1].

With respect to the glass plant, overall product separation efficiencies were low, with the resultant glass product typically high in moisture and stone content. Frequent system breakdowns occurred, and average glass plant throughputs of only 9 kg/min (20 lb/min) were achieved as compared to the design rate of 27 kg/min (60 lb/min). In addition, the glass system was quite labor intensive, requiring two to three full-time workers for its operation [1].

Of greater concern in the overall plant performance than the minor inadequacies of the pretreatment processes was that the remaining pyrolysis feed preparation units were shown to be undersized, with many only meeting half the design capacities.

At throughputs greater than or equal to 50% of design figures, the rotary kiln dryer could only decrease the waste moisture content to a minimum of 5%, as opposed to design specification of 3.5%. As throughput rates were increased, dryer abilities decreased. A problem with the malfunctioning of heat sensors in the kiln resulted in a fire when the autoignition temperature of the waste was allowed to be exceeded. In addition, occasional plugging occurred of the rotary valve feeding the dryer [1].

The rotex screen only operated at less than half design feedrates—2270 kg/hr, not 48,400 kg/hr (5000 lb/hr, not 10,655 lb/hr). Evidently, a design waste density of 80–160 kg/m³ (5–10 lb/ft³) was used in sizing the screen, although measured density was found to be 29 kg/m³ (1.8 lb/ft³). Throughput rates especially decreased when moisture content of the feed from the dryer was greater than 4% or 5%. The manufacturer of the air table was able to make adjustments to optimize its operation, although it should be noted that the unit was being fed at less than design rates [1].

Operation of the secondary shredder originally proved to be a costly and high-maintenance proposition as the attrition mill discs initially averaged only a 3-hour life span. In addition, exiting waste failed to meet design particle size distributions of nominal 1.17 mm (14 mesh). The rapid disc deterioration was thought to be caused by the glass component in the feed [1], a component that was to have been removed in previous processing steps. After some not too successful attempts to decrease glass content via alterations in upstream processing, an investigation of different materials and heat treatments with respect to disc durability was instigated. Tests on a Potamac M (cast) material gave the longest disc life—24 hours.

Like many of the preceding units, the doffing roll bin (reactor surge bin) also failed to meet design feedrates. A fire in the surge bin resulted in the adoption of a nitrogen gas purge line to the bin. In addition, a hinged door in the bottom of the bin with quick release fasteners was installed to facilitate any future fire-fighting episodes.

Pyrolysis System

As shown in Table 6.3, total time of pyrolysis system operation during the period from August 1977 to March 1978 was 140 hours. Such a small accumulation of operating time was due to the excessive mechanical problems and breakdowns experienced throughout the plant, especially in the pyrolysis system. Processing rates in this energy conversion process never

Table 6.3. Log of Pyrolysis System Operation [1]

Period	Duration (hr)	Cumulative Time (hr)
August 22-26, 1977	1.27	1.27
September 15, 1977	0.12	1.39
October 5, 1977	3.50	4.89
October 12, 1977	3.58	8.47
October 19, 1977	3.17	11.64
November 17, 1977	8.00	19.64
December 16, 1977	5.00	24.64
December 21, 1977	8.00	32.64
December 28-29, 1977	13.17	45.81
January 10-12, 1978	37.23	83.04
February 22-24, 1978	28.62	111.66
March 10-11, 1978	11.43	123.09
March 11-13, 1978	17.08	140.17

reached greater than 40% of design values [1]. Major problems in the char recirculation system were encountered, as the quantity and quality of internally produced solids necessary to replace the use of outside sources of solids for heat transfer purposes were never reached. Likewise, sufficient amounts of clean, pyrolytic recycle gas could not be produced to replace the supposedly temporary use of nitrogen gas for transport purposes and of No. 2 fuel oil for various process heating purposes.

Most importantly, properties of the small amount of pyrofuel produced did not approach product expectation. It is estimated that 11,300 liters (3000 gal) of the pyrolytic oil were produced in January and 4000 liters (1050 gal) in March. Evidently, production of fuel oil was reported "only when ORC tapped significant quantity" of the product [2]. The pyrofuel produced in other runs, according to ORC's project manager, "was primarily consumed in the solubility of pyrofuel in the quench oil" [3]. The senior project engineer during the evaluation indicated that the August through December pyrofuel productions were "contaminated with catalyst in most cases." He also estimated that the "certain quantity of quench oil" found in the pyro-oil by the evaluation was "about 20 to 25%" [2].

Table 6.4 gives a comparison of No. 6 oil and projected pyrofuel properties to the average results of six samples of pyrofuel produced at the El Cahon facility. As can be seen, product characteristics do not resemble those of a marketable fuel. Moisture content, originally predicted to be 14%, averaged 52%. Resultant volumetric heating value was only 27% of a No. 6 fuel oil, as opposed to the projected 64%.

By December of 1977, the EPA had declared the demonstration facility to be "operational" [1] and requested that an acceptance test of 72 hours

Table 6.4. Fuel Component Comparison: No. 6 Fuel Oil, Projected Pyrofuel, Actual Pyrofuel [1,8] (adapted from Refs. 1 and 8)

Parameter	No. 6 Fuel Oil	Projected Pyrofuel		Actual Pyrofuel[a]
		Dry Basis	As Marketed	
Carbon (wt %)	85.7	57.0	49.0	20.91
Hydrogen (wt %)	10.5	7.7	6.6	4.46
Sulfur (wt %)	0.7-3.5	0.2	0.18	0.13
Chlorine (wt %)		0.3	0.26	0.93
Nitrogen (wt %)	2.0	1.1	0.95	0.54
Oxygen (wt %)		33.2	28.55	18.40
Ash		0.5	0.43	2.96
Water	0.005		14.0	51.67[b]
Specific Gravity	0.98	1.30	1.26	1.312
Volumetric Heating Value (Btu/gal)	148,840	114,900	95,700	39,851
Heating Value (Btu/lb)	18,200	10,600	9,100	3,641
Pour Point (°F)	65-85		90	11.7
Viscosity (Ssu @ 190°F)	340		1150	34
Pump Temperature (°F)	115		160	As rec'd 60°F

[a] These values are averages of tests on six samples of pyrofuel produced at the El Cahon facility on December 27, 1977, January 11, 1978, January 13, 1978, January 17, 1978, April 13, 1978, and May 1, 1978. According to Chatterjee [2] the pyrofuel tests usually occurred within three weeks after the fuel had been produced. However, Holderness [3] claims the last pyrolysis run was March 13, 1978. These two statements are contradictory if the last two test dates of April and May are assumed to be correct.
[b] Moisture content of the six samples ranged from 44.4% to 58.0%.

of consecutive pyrolysis operation be met to qualify the project for further EPA financing. After five unsuccessful attempts to reach the acceptance requirements, all pyrolysis activities were halted on March 13, 1978 [3].

Significantly, the pyrolysis reactor itself had only minor operational problems, those being the plugging of the aeration nozzles that fed high-pressure gas and hot solids to the reactor. Reactor control parameter optimization was never determined, as difficulties with other process equipment precluded such a realization [1]. The particle residence time was probably too short, as suggested by the presence of large-size organics in the downstream units of the ash stripper and char burner [1].

Major problems were encountered with the solid–gas separation in the cyclones, as large amounts of solids remained entrained in the gas stream, eventually depositing in the decanter. Experimentation with variations of the cyclone design parameters of particle size and gas:solid ratios were, for the most part, unproductive. The solids seemed to be too fine to be separated by the cyclones, with attrition of the solids during their transport and recirculation promoting such fine particle size. Plugging of the dip legs exiting the cyclones also occurred, ultimately backing up the solids content in the cyclones until almost all solids were transported to the decanter [1].

The ash stripper also experienced plugging problems, requiring shutdown of the pyrolysis unit when an internal strainer became completely blocked by large organic particles and refractory from the spalling of the stripper vessel itself. Modifications giving larger strainer openings and increasing the total area for solids accumulation only prolonged the time before blockage occurred again. As in the pyrolysis reactor, clogging of the fluidization nozzles also occurred [1].

The char burner experienced occasional surges in temperature, which were to be controlled by water quenching. Complete combustion of the organic portion of the char did not occur, necessitating its combusion in the ash hopper (if it could be separated by the cyclones). The burning characteristics of the char in the hopper were unpredictable, resulting in a need for constant supervision of the unit. High- and low-temperature fluctuations necessitated use of water quenching and burning of No. 2 diesel oil for temperature control. The replacement of the existing two-layer refractory lining, which could not withstand the temperature variations, by a one-layer lining was a two-week endeavor. Flow blockage due to solids accumulation in the hopper strainer was treated with similar methods and results as in the ash stripper. In one instance, complete fracture of the dip leg feeding the hopper led to the ultimate plugging of both cyclones and the consequent transport of solids to the afterburner. Replacement of the dip leg was difficult as original design did not include easy access provisions for such an operation [1].

Problems in the operation of the ash elutriator were experienced when the level control malfunctioned. The consequential inability to correctly empty collected ash into the storage vessel required at least three men to manually unload the elutriator. Frequent seizing of the valve between the ash hopper and elutriator also occurred.

As insufficient quantities of ash were collected in the ash circulation system, the function of the ash storage vessel was subsequently changed into a storage vessel for the solids initially projected to be needed for startup conditions only. Estimates of 27 metric-tons (30 tons) of the spent fluid cracking catalyst were needed for startup conditions, with periodic replenishment required to compensate for the estimated 0.4–1.8 metric-tons (0.5–2.0 tons) of catalyst lost to the decanter and baghouses each hour.

Charging of the catalyst to the hopper was done by fluidization with high-pressure gas. In one instance, fluidizing pressures were so high that a pressure-relief device allowed a major spill and loss of 2–3 tons of catalyst to the atmosphere in a matter of minutes [1,3]. The evaluation report claimed the occurrence to be "due to the negligence of the operators," however [1]. Alterations were subsequently taken so any further pressure relief measures would direct solids to a dump storage bin.

Because the catalyst was expensive ($220–$300/metric-ton; $200–$270/ton) and difficult to obtain (several startups were delayed because of lack of catalyst supply), experiments with other solids were begun to find a less expensive and more available heat transfer medium. Initial experiments with F-140 Ottowa sand indicated that higher aeration rates for the sand were needed due to its higher density and that an air heater fired by No. 2 diesel oil was needed in the hopper to make up for the sand's lack of catalytic action [1]. Supposedly, more stable circulation was achieved with the sand [3].

The solids not captured in the char recirculation system deleteriously affected the recovery of the remaining pyrolysis products. None of the downstream processing units were designed to handle the high solids component in the pyrolytic vapors, and the resulting processing inefficiences of each unit had a domino-like effect on the following unit's capabilities.

Solids accumulation in the quench venturi spill chamber and weir mouth diminished the quenching action of the No. 2 diesel oil on the pyrolytic vapors. Several times this buildup caused the quench oil to enter the gas flow pipeline and, ultimately, the afterburner unit. Basically unsuccessful attempts in process improvement included the increase of oil flowrates from 3.5 liter/sec (55 gal/min) to 22 liter/sec (350 gal/min) for improved quenching abilities and the relocation of a quench-oil temperature sensing device to provide better temperature control [1].

Affected by the inadequate quench operation, the resulting poor quality of pyrofuel, the poor operation control procedures and, most importantly, the high solids content of the incoming gas, the decanter operation did not meet design expectations. The inevitable transformation of the pyrofuel decanter boot into a solids trap forced the relocation of the pyrofuel pump to the bottom of the decanter for the collection of whatever pyrofuel was produced. Many decanter coalescer pads were permanently removed after plugging by the solids. Removal of the solids by outside contractors was a costly, one-week proposition. In addition, the carryover of pyrofuel as aerosal mists was increased by the accumulating solids [1].

Occidental attempted to make up for the decanter's inabilities by developing the "Bakertank," a secondary sedimentation system with baffle plates for the separation of pyrofuel, quench oil and solids. The solids in the decanter would not transfer to the new tank, and the attempt to salvage the unit process was deemed a "complete failure" in the evaluation report [1].

Acceptance of the pyrolytic gas with aerosol mist and entrained particles caused great solids accumulation (mostly tars and organics) on the tellerette plastic packing rings in the acid scrubber. The demister mechanism functioned poorly and continued to do so even after attempts at improvement. A bar spray system was subsequently installed for the gas cleaning.

The low-pressure, high-speed, single-stage centrifugal compressor was dubbed "the most troublesome piece of equipment in the pyrolysis process" [1]. The 220-kW (300 hp), 35,000-rpm unit was designed to accept gases of $54°C$ ($130°F$), 35 kPa (5 psig) and produce gases of $180°C$ ($350°F$), 207 kPa (30 psig) [1,3]. The presence of the aerosols and solids, however, threw the compressor off balance and caused failure of impeller blades, impeller shaft and the shaft seal, rendering the unit inoperable in roughly 3–4 hours. Results of such improvements as the installation of drainlines for capture of tar-like liquids, the adoption of centrifugal aerosol separators for capture of droplets and solids less than 10 μ and the use of filter beds for capture of greater than 3-μ matter, increased compressor life to 10-12 hours [1].

The high-pressure compressor, designed to produce a gas of $230°C$ ($450°F$) and 308 kPa (45 psig), had similar, but less severe, problems as the low-pressure unit because many of the solids and aerosols had deposited in the low-pressure compressor.

It is interesting to note that backup compressors were available during operational activities, the only real example of redundancy in the entire facility. As this redundancy was not mentioned in project projections, such a feature was probably added after initial problems were encountered. Electrical switching equipment was installed between both the high- and low-pressure compressor standbys to allow switchover without loss of recycle gas [1]. Such an addition would undoubtedly help increase the online time during the EPA acceptance tests.

The operation of the afterburner proved to be expensive as No. 2 diesel oil was used as the fuel source because sufficient quantity and quality of recycle gas and pyrofuel were not available. Addition of an automatic pressure control valve was adopted during plant operation for use in maintaining pressures so flameouts would not occur in the combustion chamber. Affected by the high solids content of incoming gas streams and the stop-and-go operation of the plant, the refractory lining of the combustion chamber frequently needed repair or replacement. In one instance, an extremely high solids input resulting from the previously mentioned fracture of the ash hopper dip leg caused complete refractory erosion. Another time, a combination of operator error and control and equipment malfunctioning allowed quench oil to enter the combustion chamber. The ensuing rapid temperature rise caused warping of the heat exchanger tubes. The flue gas valve and pipeline system, also not designed for the high solids content, experienced heavy refractory erosion that caused shutdowns and necessitated costly and often lengthy repair.

Occasions of high solids loading to the dryer vent (afterburner) baghouses could not be handled adequately by the as-designed devices, resulting in very dusty conditions around the baghouse area [1]. At one point during facility operation, the spray nozzles cooling the afterburner malfunctioned and consequently oversaturated the gas stream traveling to the baghouse. Plugging of the bags, problems in the rotary valve of the baghouse and severe condensation in the baghouse ensued. The troublesome nozzles were subsequently replaced with mist-type nozzles, albeit with marginal results.

Successes with the remaining baghouses were much greater. Visual stack examination usually showed a clear stack. Only during times of high catalyst carryover during "system upsets" did the afterburner baghouses become inadequate for particulate pollution control [1].

High noise levels coming from the afterburner baghouse blower, especially apparent at night when background noise was minimal for the area, annoyed nearby residents. Results of the addition of sound-deadening materials were inconclusive as plant shutdown occurred soon after. Another major noise contributor—the rodmill in the glass recovery system—was successfully muffled via enclosure by sound-deadening materials. The liquid impoundment pond was continuously overloaded from the frequent plant shutdowns. Outside vendors were brought in to clean it, removing the accumulation of spent catalyst and tarry solids as well as the oily and greasy material floating on the top of the pond [1].

EVALUATION

A hindsight evaluation of the San Diego demonstration facility shows design inadequacies to be the most significant factors behind the apparent failure of the flash pyrolysis system. Failure, for the sake of this discussion, is judged by the designer's termination of further activities at the El Cahon pyrolysis site as of this writing, in conjunction with its continued involvement in other types of waste-to-energy projects in the country.* In addition, the other two financial backers of the facility (the EPA and San Diego County) did not feel the plant worthy of additional investment to enable its operation. Two nontechnical and less significant points should be discussed before a more technical look at the pyrolysis system is given, however.

First, the evaluation report and the above conclusions are based on "short-term, non-steady state, reduced capacity runs" [1]. Few of the major equipment items were at full design load conditions during data collection. Upstream inadequacies were constantly passed on to the downstream processing units, making their original designs inadequate for processing the "undesigned-for" feed characteristics. Difficulties in meeting the EPA acceptance test of pyrolysis operation were often due to specific unit inabilities, not system inabilities. A "failure" classification for the entire flash pyrolysis system could therefore be too general a verdict. This leads to the second point.

One might argue that an experimental facility of such a multiunit, interdependent and nonredundant nature as the San Diego plant warrants a long shakedown period. Total plant operation, including shakedown, was for a little more than 1.5 years. The pyrolysis system was only operated over half this time period. Such short operational time before the closing of a plant may be unfair to the system.

The decision to close the facility after such a relatively short operational time may also have reflected the situation in San Diego County. Unlike Baltimore City, San Diego County could easily afford to landfill the extra tonnages of solid waste not processed at the facility. It was not necessary, therefore, for the municipality to try to make the facility work, as Baltimore did when Monsanto pulled out of the city's solid waste demonstration facility. Interestingly, however, San Diego County is still responsible for the lease payments of the site until April 1, 1983, as specified in a five-year contract with Occidental. To help defray such ownership costs, the county is negotiating with several firms interested in using the facility and/or site in some form. ORC maintains "the right of first refusal" in operation of the plant under this contract, however [5].

Pyrolysis System Evaluation

The pyrolysis of waste, involving its "irreversible chemical change brought about by the action of heat in an atmosphere devoid of oxygen"

---

*Interestingly, as of fall 1979, Occidental has withdrawn from actively pursuing resource recovery proposal requests.

[18], produces products of char, organic liquid, fuel gas and water. The actual percentages of the product breakdowns and their compositions are determined by many process parameters, including waste feed characteristics, pyrolysis temperature and pyrolysis residence time. As general product output trends are associated with particular parameter values, a technical evaluation of the pyrolysis system necessitates a look at product expectations for the facility. In the case of the San Diego demonstration facility, the desired product was a marketable fuel oil of 14% water. Char and pyrolytic gas by-products, also produced during the pyrolysis of waste, were to be used for plant processing.

The design temperature of 510°C (950°F) for the pyrolysis reactor fits in with the desired product results. Generally, pyrolysis temperatures on the order of 500°-540°C (930°-1000°F) produce high yields of pyrolytic oil, while temperatures on the range of 700°-1000°C (1290°-1830°F) produce primarily gaseous fuels. Pressures used for both of these processes are roughly atmospheric [19,20].

The characteristics of the pyrolysis feed and some of the unit design processes used in the San Diego project also follow general guidelines for optimization of liquid fuel formation. The yield of liquid organics is known to increase when waste is *rapidly* heated at the previously mentioned temperature range and then *immediately* quenched below the pyrolytic gas dewpoint to condense the fuel fraction from these gases and vapors [19]. Secondary shredding in the Occidental facility produced a feed of small particle size, improving overall heat transfer efficiencies in the reactor and ultimately reducing the required residence time of the feed in the reactor [8,19]. Further, the unit processes following the gas–solid separation of the pyrolysis products involve the immediate quenching of the gas and vapor product.

Occidental's feed-drying step, besides improving the functioning of some of the feed preparation equipment and decreasing the heat needed for the pyrolysis reactor, helped to improve the quality of the liquid fuel product by decreasing its resultant moisture content. Although a 14% water content was desired for the pyrofuel, higher percentages of water would decrease its heating value and, therefore, its marketability. As total water output from the pyrolysis reaction is the sum of water by-products produced from the reaction and of water already in the feed, a dry pyrolysis feed would consequently help minimize the water content in the pyrolysis products [19,20]. In addition, it has been claimed that the actual amount of the water output found in the pyrolytic oil product is largely a function of temperatures maintained when condensing the pyrolytic gases [3].

Another pyrolysis feed processing step used in the San Diego plant helped improve the quality of the char product. Removal of most of the inorganics before pyrolysis would ultimately decrease the ash product of the char [20], thus increasing the ability of the char to function as an indirect heat transfer medium. It should also be noted that other parameters, such as increased pressure, addition of catalysts and usage of oxidizing and reducing reactants, can influence ultimate product breakdowns.

Unfortunately, incorporation of the mentioned parameters into a facility design does not guarantee the production of the expected outputs, as experience shows. To begin with, authorities in the field of pyrolysis do not hesitate to point out the uncertainties of the process, even though generalities are known. After looking at studies of pyrolysis process parameters, Lewis [18] concluded that "the complexity of these interactions makes it impossible to predict the final product characteristics." Jones further stated that [19]:

> Presently, it is difficult to predict the proper operating conditions that achieve product yields for most heterogeneous waste materials having variable moisture and ash content. Empirical methods must be used.

Bench-scale and pilot-sized models are supposedly used to provide just such empirical data. However, from experiences of both the Baltimore and San Diego pyrolysis demonstrations, it is apparent that empirical data gained from such small versions of a specific pyrolysis system cannot always be used successfully in the scaleup of such a process. Perhaps more gradual scaleups of such a complex process as pyrolysis are warranted.*

Holderness feels that a majority of the solid separation problems in San Diego "were the results of extrapolation or sizing problems from the research facility" [3]. Discounting the relatively minor problems with specific unit breakdowns or malfunctions, the continual difficulties with solids separation seem to have been a major cause of pyrolysis system technical and financial inadequacies.

Besides the inherent uncertainties in the pyrolysis process and the presence of scaleup difficulties, several other factors increased the possibility of failure with respect to original expectations in the San Diego process. According to a reliable source, who did not wish to be quoted, the 3.6 metric-ton/day (4 ton/day) pilot facility did not include either char recirculation or gas recirculation—two major system components by which the economic viability of the overall project was greatly influenced.

Without the designed quantity and quality of recovered char and pyrolytic gas, operational costs would increase significantly as more nitrogen gas, No. 2 fuel oil and startup solids would be required. Had

---

*An interesting view toward the scaleup of a pyrolysis facility was taken by a Danish company called Pollution Control Ltd. After successfully proving a 5 metric ton/day plant of the "Destrugas" process, plans for larger facilities of 100–500 metric ton/day included banks of the proven, smaller sized units. It was felt the heat transfer characteristics would not be maintained otherwise [21].

actual operational problems with the systems been known, designs might have incorporated onsite nitrogen gas production, for example, to help decrease costs [1], although No. 2 fuel oil and a solid-circulation media would still be necessary.

As with the Baltimore facility, the failure to include all processing components of the larger facility in the smaller scale models resulted in serious problems. Such an omission almost guarantees the need for additional shakedown time and modification costs. It can also prevent a realistic outlook of the potential economic or technical viability of a project.

The complex and basically nonredundant design of the San Diego facility inherently reduced the chances of its functioning in a smooth or continuous fashion. As pointed out before, specific unit breakdowns frequently stopped the entire operation of the facility. Further, malfunctioning of upstream units affected the downstream units' capabilities. The 1979 evaluation report claims that limited funding led to "stop-gap measures" [1]. Realistically, however, an unproven design requires an initially *high* contingency fund for startup operations and modifications, along with a patient attitude toward shakedown operations.

The report also started that "the lack of well-trained operators in this complex and highly instrumental system [the San Diego facility] resulted in many operational errors and caused frequent shutdowns" [1]. Such a statement refers to operational, not technical, inadequacies. However, if operational problems were as frequent as insinuated in the report, much money must have been invested in correcting any technical problems caused by operator errors—money that could have been spent on equipment or design modifications. Moreover, these problems would indicate insufficient training of personnel and/or too complex a system without adequate control or feedback mechanisms.

Questions still remain and points are still unclear as to what actually happened to create such a drastically different liquid fuel from the demonstration plant as compared to the pilot facility. Known factors contributing to such an inferior quality product include the poor quenching process due to solids interference; the poor collection process due to solids interference in the decanter; and the sometimes high moisture content of the pyrolysis feed due to insufficient rotary kiln operation. Factors possibly contributing to a poor quality product include inconsistent and perhaps inadequate heat transfer in the pyrolysis reactor due to solids circulation problems and too long a time period between pyrolysis and quenching of the pyrolytic gases (resulting in further decomposition of oil precursors than was wanted).

Storage of the produced fuel resulted in separation of some of the water component from the pyrolytic liquid due to the liquid's higher specific gravity [2]. In addition, it was felt that no chemical changes occurred in the liquid during storage [2]. Pyrolytic liquid properties after storage (and consequently after some water removal) are not available, nor is it known whether tests such as those were made. However, it is believed that if the liquid produced had a potential value as a fuel once water separation had occurred, such a fact would be publicly known.

The evaluation report justifiably did not try to project net costs per ton for larger facilities similar to the San Diego project "because the flash pyrolysis was not proven and there is insufficient data from which to make 'scale-up' projections." Similarly, environmental evaluations of the flash pyrolysis process were not attempted because steady-state conditions were not met, design loads were not processed and operations were usually not long enough to perform such testing [1].

Perhaps the best summation of the San Diego pyrolysis facility results, besides the rhyme at the opening of the chapter, can be seen in the answer to the question, "If you were to design the facility over again, what might you change?" Two Occidental employees indicated they would change the ultimate product breakdown, probably by altering the pyrolysis temperature, to optimize the pyrolytic gas production. In this day of skyrocketing oil costs, one would think that if a process could produce an oil from MSW, it would be pursued. Evidently, oil prices will have to increase much more before an organization will take up the flash pyrolysis process where Occidental Research Corporation left off. When that day comes, and it may come soon, we will have Occidental, as well as EPA and San Diego County, to thank for their good faith effort that ultimately led to the collection of invaluable empirical data on the possibilities of producing a fuel oil from municipal solid waste.

REFERENCES

1. Acres American Incorporated. "Technical, Environmental and Economic Evaluation of the San Diego County Resource Recovery Facility at El Cahon, California," report prepared for Office of Solid Waste Management, U.S. EPA, Contract No. 68-01-4420, Raleigh, NC (1979).
2. Chatterjee, A. K., On Site Project Monitor for Acres America Inc., during the evaluation of the pyrolysis facility. Personal communication (1979).
3. Holderness, E. R. Hooker Specialty Chemicals Division, Niagara Falls, NY. Personal communication (1979).
4. Resource Planning Associates. "Financial Methods for Solid Waste Facilities," EPA Publication PB 234-612, (Springfield, VA: National Technical Information Service, 1974), pp. 344-345.
5. Clay, N. A. Chief of Solid Waste Operations, San Diego Department of Sanitation and Flood Control. Personal communication (1979).
6. Levy, S. J. "San Diego County Demonstrates Pyrolysis of Solid Waste," EPA Publication SW-80.d.2, U.S. Government Printing Office, Washington, DC (1975).

7. Ralph M. Parsons Company. "Engineering and Economic Analysis of Waste to Energy Systems," EPA Publication 600/7-78-086, (Springfield, VA: National Technical Information Service, 1978), pp. 227-44.
8. Preston, G. T. "Resource Recovery and Flash Pyrolysis of Municipal Refuse," *Waste Age* 5:83-89 (1976).
9. Mallan, G. M. "Flash Pyrolysis Turns Refuse to Fuel Oil," *Chem. Eng.* 15:90-91 (1976).
10. *San Diego Tribune* (August 1, 1978).
11. *San Diego Union* (September 13, 1978).
12. Garbe, Y. M., environmental engineer, U.S. EPA. Personal communication (1979).
13. Chatterjee, A. K., and Y. M. Garbe. "An Overview of San Diego County's Resource Recovery Plant," *Proc. 1978 Nat. Waste Processing Conf.*, Chicago, IL, (1978), pp. 447-55.
14. Garbe, Y. M. "Demonstration of Pyrolysis and Materials Recovery in San Diego, California," *Waste Age* 5 (1976).
15. Boegly, W. J., W. R. Mixon, C. Dean and D. J. Lizdas, "Solid Waste Utilization-Pyrolysis," Oak Ridge National Laboratory, Oak Ridge, TN (1977).
16. *San Diego Tribune* (March 24, 1978).
17. *San Diego Union* (January 14; July 18, 1978).
18. Lewis, J. M. "Thermodynamic Fundamentals for the Pyrolysis of Refuse," Proceedings of the National Waste Processing Conference, Boston, MA (1976), pp. 19-40.
19. Jones, J. "Converting Solid Wastes and Residues to Fuel," *Chem. Eng.* 1:87-95 (1978).
20. Weinstein, N. J., and R. Charanjit. "Pyrolysis/State of the Art," *Public Works* 4:83-86 (1975).
21. Conn, W. D. "European Developments in the Recovery of Energy and Materials from Municipal Solid Waste," U.S. EPA Publication 600/7-77-040 Springfield, VA: National Technical Information Service (1977).

CHAPTER 7

CONCLUSION

The previous chapters deal with the various phases in the conception, planning and operation of five high-technology resource recovery systems. Each recovery facility applied, or was to apply, a different technology, ranging from the uncomplicated design of mass incineration at Nashville Thermal to the complex design of the San Diego flash pyrolysis facility. In addition, each of these projects experienced failure in the sense that original technical and/or financial expectations were not realized.

Although common denominators affecting the relative successes or failures of these facilities can be found, *many* factors contributed to each facility's outcome, and these factors varied in importance for each system. Therefore, besides learning from the specific mistakes and successes mentioned in the case studies, it is just as important that one major conclusion be reached: every recovery facility will be affected by many parameters, each having different importance to a project's success as determined by local and/or project specific conditions. When planning for recovery systems, then, all possible contributing factors to a project's success should be addressed, no matter what the system. The wide range in factors and their effect on the projects studied point to such a conclusion.

COMMON DENOMINATORS INFLUENCING FACILITY SUCCESS

Many factors contributed to the outcome of initial operations of the five facilities studied. The most significant and common parameters are found in Table 7-1. Such a list is by no means all-inclusive, however, as other recovery

Table 7-1. Major Parameters Affecting Recovery Facility Success

Area Markets	Alternate Disposal Possibilities
Initial Funding	• Cost
Contingency Funding	• Location
Commitment for Success	• Waste flow control
Technical Viability	Time Pressure
Waste Flow Control	

attempts may be strongly affected by parameters not common to the studied facilities.

One of the most important factors affecting a recovery facility's success is the presence of markets for the recovered product. The Nashville facility survived because of an assured market for its product—steam. The markets for incinerator residue at Lowell and Baltimore were soft or nonexistent. The Occidental facility could not sell its products, however, as the pyrofuel did not meet product standards as originally claimed.

Perhaps the most obvious factor influencing the success of a facility is the available funding behind a project. Nashville Thermal's experience was largely unsuccessful when insufficient funding was available. Likewise, Baltimore was unable to make needed modifications until a grant was obtained from EDA, although it has not yet been determined whether these changes have been wholly successful. Union Electric's inability to obtain adequate funding for its proposed Solid Waste Utilization System resulted in the abandonment of the proposal. Lowell was not willing to invest more money in its incinerator to pursue the recovery project. Of the five projects, only Occidental Research Corporation did not have funding problems. Instead, it decided that investing more money into the system would not be worthwhile. The lesson to be learned is that a significant contingency fund is necessary in the implementation of a resource recovery facility. All the facilities investigated required unexpected alterations of one form or other during shakedown: Nashville had to find and correct leaks in the district systems as well as provide services using fossil fuels; the St. Louis demonstration facility added an air classifier and tested material for the pneumatic piping; Baltimore experienced unexpected occurrences of minute particles, necessitating more expensive pollution control equipment; and San Diego experienced major problems with gas—solid separation, adding to downstream unit processes' maintenance costs. Sufficient shakedown funds should be budgeted in the original project financing to enhance the success of the project.

Another common criterion affecting a facility's success is the commitment of the sponsoring agency or organization. In both the Baltimore and Nashville experiences, the cities were relying on the successful operation of their respective facilities. Both desperately needed additional waste disposal, and Nashville was dependent on Thermal's services for heating and cooling some of its buildings. Although Bi-State knew that disposal problems were imminent for the city of St. Louis, it knew the city had other, more realistic options to consider than just a version of the Union Electric proposal. Most importantly, the agency was able to cost-out a large-scale facility not yet built before deciding whether to continue with the original proposal. Nashville and Baltimore did not enjoy that luxury. San Diego County felt the least commitment of all to make its facility operate as its area disposal alternatives were, and are, sufficient. Significantly, the two facilities having the most committed backing currently seem to be the most successful of the five.

The technical viability of the processes used in the four EPA demonstration grants had never been determined in a full-scale system. San Diego and Baltimore, the most unproven of the five, were strongly affected by the technical inabilities of the large systems. In Baltimore's case, modifications have been undertaken to correct the technical deficiencies. The Occidental pyrolysis design, however, currently seems inoperable. The major problems experienced by these two facilities are inherent to the EPA demonstration grant process that required full-scale, unproven technologies to obtain funding assistance. The St. Louis demonstration was designed to minimize potential technical problems by using limited processing and by changing the conditions in the coal-fired boilers as little as possible. Although Nashville had some technical problems, these could have been largely avoided if original design had combined state-of-the-art knowledge.

Waste flow control (or lack of) is another common factor, although its appearance only stood out in St. Louis and Lowell. Had Union Electric been able to get all its transfer stations and had Proposition 1 never passed, the utility may have been able to establish a viable system *if* a way of securing the 7200 metric-ton/day (8000 ton/day) of solid waste had been determined. Lowell's inability to capture initial residue tonnage estimates from surrounding areas was unique, as this was dependent on the towns utilizing their own incinerators. Waste flow control may not have legally worked in this case. The Nashville facility would have been detrimentally affected if METRO had not been able to direct solid waste to the facility when Thermal began charging tipping fees. Likewise, as Baltimore's operation becomes more stable, the importance of having the city collect and deliver wastes should be more apparent. San Diego had a unique situation in that a local private hauler delivered MSW to the pyrolysis plant without incurring a tipping fee.

Availability of other competitive disposal alternatives with respect to cost and convenience can determine whether a proposed system will stand a chance of successful operation. The decision to terminate the St. Louis Solid Waste Utilization System, for instance, was made when it was realized the system was not competitive with area landfills. This factor does not seem so important, however, when waste flow control is not an issue. For instance, both Nashville and Baltimore could direct their municipal haulers to their respective facilities even if said facilities were not economically competitive or conveniently located.

Although not as significant as the other parameters, the absence of an extreme time pressure on a facility's operation can improve a system's chance of successful operation. The Nashville Thermal facility's timely operation was so important that spending additional time to obtain more funding was thought to be unfeasible. Additionally, the Nashville facility began operation before construction was complete, adding to its operational costs. Even a large amount of publicity before a project's initiation can add pressure to a facility in meeting operational status before an adequate shakedown period can be completed. It appears that both Baltimore and Nashville felt these pressures.

CHANGES IN RESOURCE RECOVERY PLANNING

The common denominators influencing a recovery facility's success as found for the five systems studied pertain to present-day recovery planning. Other considerations are continually being added to the checklist of factors to consider as experiences in resource recovery are made public.

The history of the five recovery systems shows the need for extensive planning before project implementation. Such realistic attitudes toward resource recovery are developing; and communities are not jumping into a recovery plan without thorough background investigations, such as determining needs for waste flow control and determining competitive solid waste disposal alternatives. Priorities have changed to those of securing an environmental, economically competitive, convenient way of disposing of solid wastes, as opposed to the recovery of lost resources in the waste stream.

The field of solid waste consulting has developed with the realization that many factors must be considered when planning a recovery facility. In the case of the four demonstration grant facilities examined, each facility was linked directly to specific technologies without carrying out an in-depth study to determine whether such a system might be viable or whether the system was the best for the area. Interestingly, once a solid waste consultant was hired to study the St. Louis situation, it was determined that any form of the Union Electric proposal would not be a feasible solution for the city's growing solid waste disposal problems.

CONCLUDING REMARKS

In looking at the factors affecting the successes and failures of the five facilities examined in this book, it is apparent that many factors ultimately contribute to a facility's outcome. All five projects were initiated during the infancy of the resource recovery push in this country, without the advantage of the realistic outlook in planning that is evident today. Additionally, four of the facilities were a result of the EPA demonstration grant funding, which further decreased the parameters considered (technical options were already chosen, for example) in the planning of the facilities.

Complete knowledge of past experiences in resource recovery attempts is invaluable in current and future recovery planning. Presentation of the facts behind the five case studies in this book should aid in furthering such knowledge.

INDEX

afterburner 70,77,78
air classifier 24,68,72
Air Conditioning Corp. 13
air quality 14,15,17,20,21,37,39,41,
  42,44,48-59,63,70,79
air table 73
aluminum 48,49,62-66,71,72
American Air Filter Co. 17
American Society for Testing and
  Materials 2
area markets 88
Arthur Anderson & Co. 17
ash 37,69,70
Avers, C. 19

Babcock & Wilcox 10
baghouse 77,78
Bakertank 78
Baltimore Gas & Electric Co. 37,39
Baltimore, MD 35-45,88-90
Bi-State Development Agency 29-33
boiler tubes 16,17
bonding 8,11,12-17,29,55
Briley, Mayor Beverly 9

Carrier Corp. 10
Chambliss, C. 17,20
Chapman, D. 41
Chelmsford 55

coal liquefaction 62
collection of refuse 9,27,31
Combustion Engineering 6,10,24
compressors 78
conveyors 39,66,71,72
corrosion of boiler tubes 16
cracking 69
cyclones 69,76

demonstration grant 13,24,36,54,
  56,61
Detroit Stoker Co. 10
doffing roll bin 66,72,73
Dougherty, C. 29
dryer 68,70,73
Duff and Phelps, Inc. 13

Economic Development Administration 42
eddy current separator 68
editorials (in the press) 15,47,55
Ehrhart, Div. of Procon, Inc. 63
Eigner, J. 30
El Cahon, CA 63
electromagnet 66,72
electrostatic precipitators 13-15,24,
  42
Energy, Department of 2
Engineering Society of Baltimore 42